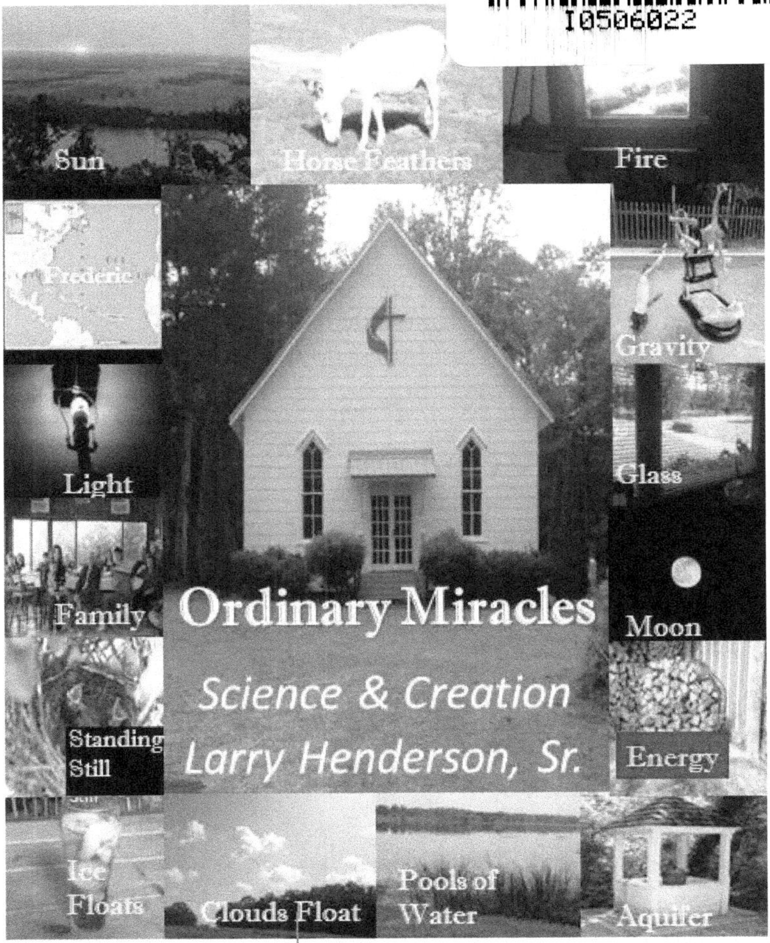

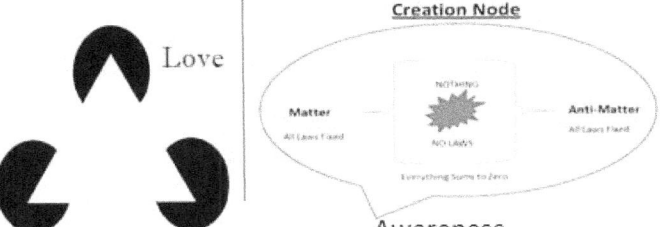

Sun
Horse Feathers
Fire
Frederic
Gravity
Light
Glass
Family
Moon
Standing Still
Energy
Ice Floats
Clouds Float
Pools of Water
Aquifer
Ordinary Miracles
Science & Creation
Larry Henderson, Sr.

Love
Creation Node
Matter
All Laws Fixed
NOTHING
NO LAWS
Anti-Matter
All Laws Fixed
Everything Sums to Zero
Awareness

Ordinary Miracles
Science & Creation

ISBN-13: 978-1493753482 (CreateSpace-Assigned)
ISBN-10: 1493753487
Copyright © 2013 by L Henderson. All Rights reserved

Published by L Henderson
1105 Moody Rd.
Lucedale, MS 39452
Email: l_henderson1@bellsouth.net

Printed by CreateSpace
Available: Amazon, Ingram, Baker & Taylor,
ordinarymiracles.us and bookstores everywhere

This book, or parts thereof, may not be reproduced, stored in a retrieval system, or transmitted in any form or by any means, (electronic, mechanical, photocopying, recording or otherwise), without the written permission of the publisher.

Credits
Every attempt has been made to credit the sources of copyrighted material used in this book. If any such acknowledgment has been inadvertently omitted, receipt of such information would be appreciated. Wikipedia is the verification source of all data.

Ordinary Miracles

Science & Creation

Larry A. Henderson, Sr.

An
L Henderson Publication

Ordinary Miracles

Contents

Acknowledgements	vii
Foreword (status quo)	ix
Family	1
Hurricanes	7
Standing Still	9
Ice Floats	13
Which is heavier dry air or wet air?	15
Snowflakes	19
Cycles	21
Day and Night	25
Nitrogen Atmosphere	27
Nitrogen Cycle	29
Energy Availability	31
The Anomaly of Oxygen (availability as O_2)	33
The Anomaly of Oxygen (abundance)	36
Pools of Water	37
The Well	39
Running Water Uphill	41
Water Anomalies	43
Mean Global Surface Temperature	47
Ecosystem Balance	49
The Miracle of Gravity	53
Inspiration	55
The Miracle of Words	59
Plants and Animals Made of the Same Stuff	63

Horse Feathers	65
Vegetables	67
Seeds	69
Smart Seeds	71
Makers	73
Glass	75
Teachers	77
Leadership	79
Retirement	81
Light is Something Else	83
Fire	85
The Sun	87
The Moon	89
A Tilted Earth	91
Magnetosphere	93
The Ozone Layer	94
The Human Brain (hardware)	95
The Human Brain (software)	97
Evolution	99
Genetic Engineering	100
The Speed of an Electron	103
Enzymes	104
Hemoglobin	105
Glucose/Ketones	106
The Eye	107

Artificial Intelligence	109
Love	111
Spider Web	115
Life	117
Integrity/Relativity	120
Antarctica	124
Who is God and where did He come from?	125
Conclusion	129
Bio and Ordinary Miracles	131

Acknowledgments

Thanks to my wife, Ann, for her no-holding-back input. She did not try to protect my feelings or ego with her always on-the-money feedback. Her sobering comments helped keep the work true and honest.

To my children and major source of material for life itself, I am thankful for their volumes of information. Larry Jr. kept me straight on any scientific issues that I may have strayed on as well as being a good sounding board. His wife Hallie did a good job of editing and providing grammatical counseling. My daughter Leigh Ann, as always, is the solid rock of support in this endeavor. She pointed out several ordinary miracles and contributed so much in photography and editing the manuscript. Leigh Ann's husband Matthew Cashwell has been a solid contributor with ordinary miracle ideas and a good weekly feedback source.

Bro. Danny Box, Bro, Gene Vance (my pastor), and Bro. Cecil Locke provided spiritual input, material, and editorial assistance. Danny's mother Mary Box also read the manuscript and found an error. Of all the people that read the document, Mary alone noticed that in reference to the mammary gland of a hog I had used utter instead of udder. I appreciate this find. It would be embarrassing if my readers found that I had such uncorrected ignorance. Thanks, Mary.

Thanks to Sue Fountain, Evie McNeese, and Sue Crocker for reading and enriching developing manuscripts.

Special thanks go to my neighbor Melba Futral for her review and constructive input to the manuscript. She and her husband Marlin are rock solid people that anyone would be proud to have as neighbors.

This book would still be in the form of Sunday morning notes without the input and encouragement of my book buddy, Ann Smith Hill, whose editing, helpful advice, and belief in this work are invaluable. Ann is author of the book *God Moment Miracles* subtitled: *Life with Muscular Dystrophy*. Her book is a trip down memory lane of her life dealing with the loss of her husband and children to this deadly disease. Ann's book is energizing and gives positive encouragement for dealing with life's complications.

Thanks to my grandson Christopher Johnson for contributing the section on Antarctica and generating some enjoyable discussions. My other grandchildren, in order of age, Luke Cashwell, Oren Cashwell, Bailey Henderson, and Jordan Henderson contributed with lessons on a daily basis of what life is all about.

Thank you, Ina Baygents, for being the best mother-in-law and general supporter that a man could possibly have.

Wikipedia is the verification source of all data in this book.

Foreword

Life on Earth is surrounded by many extraordinary things that we simply take for granted. There are numerous improbable scientific phenomena that strongly suggest a design influence. This writing describes fifty-two (52) design-influenced topics: miracles we see every day that science cannot explain. Ordinary miracles are things we consider normal that are outside reasonable expectations of occurrence unless a design influence is present. Science is a powerful tool in understanding the beauty of design. Miracles are not exceptional events outside of "natural laws" but rather they demonstrate natural laws are part of an overall design. The systems that are essential for our existence and exceed the probability of random chance occurrence are abundant. The absence of even one of these life-supporting systems and we could not exist.

The probability of an occurrence is the product of probabilities of all of the independent subcomponents. Suppose you have a box of beads where half of them are red and the other half black. The probability of drawing a red bead on the first draw is ½. The probability of drawing a red bead on the second draw is also ½ if replacement of the first bead was made in order to keep the mix constant. However, the probability of continuing to draw consecutive reds is the product of the probability for each draw. Therefore, drawing two consecutive reds is ½ x ½, which is $(½)^2$ or ¼. Drawing 52 consecutive reds is $(½)^{52}$ or

1/(4,503,599,627,370,496). The odds go down rapidly as subcomponents are added. Keep in mind this is an example where probability is based on a fifty-fifty chance. The probability of Earth ending up in its current form and with all of its physical properties is the product of a multitude of very-low-probability events. Consider the likelihood of a large body hitting the earth during early formation and knocking the planet off perpendicular by 23 degrees while generating a stabilizing moon ¼ the diameter of the earth. This is less likely by several orders of magnitude, and this is only one subcomponent of Earth evolving into the earth we know. Get the product of several hundred of these low probability events, and you have a probability of 'It did not happen without design influence.' The probability ratio surrounding light, which is explained in the section *'Light is Something Else,'* approaches zero.

Warren Buffett and Quicken Loans offered a billion dollar prize for correctly filling out the 'March Madness' basketball brackets. Mathematicians and statistics savvy people pointed out that they were really not at risk due to the impossibly low odds. To win the contest, one must correctly pick the winner in all 63 games. If the games had a probability of 0.5 or 50% for each game the odds of winning the Grand prize is 1 in 2^{63} or 1:9,223,372,036,854,775,808. If everyone in the U.S. filled out a bracket at random, you could run the contest for 290 million years, and there would still be a 99 percent chance that no one would win. Now, while the random chance evolutionist will admit that this is impossible they have no issue with the earth just happening even though the odds are orders of magnitude worse.

There are many laws of science of which we have empirical knowledge, not designer understanding, such as gravity. We know how to measure the effects of gravity and how to use gravity; however, we have no clue of what generates the force. The same is true for magnetism and electricity. Many ordinary miracles are such that we fully understand the phenomenon's existence; however, we are not able to explain the rationale that makes it possible.

I was chided by a friend, "Based on the contents of your document, it seems that you really enjoy musings about life and nature. So, challenge yourself to ponder the "wonders" of non-anthropogenic cancer, Parkinson's disease, muscular dystrophy, genocide, birth defects, child abuse, Rett syndrome, slavery, Lou Gehrig's disease, schizophrenia, etc." One possible answer is that all of the mentioned maladies may be anthropogenic (resulting from the influence humans have on the natural world). I do not pretend to understand.

What is the prevailing perception of our knowledge of science? How much of science is still unknown, not by an individual, but collectively? The truth is, we cannot know how much is remaining to be known. You cannot know what you do not know. Charles H. Duell, the

Commissioner of the US Patient Office in 1899, is accredited with saying, "Everything that can be invented has been invented." And although the literal quote cannot be verified, the context was stated in a report to Congress by his predecessor, Henry Ellsworth. This describes the high opinion of our intellectual achievement held by past and current humanity. We are so impressed with ourselves due to the intelligence we see on display. 'This has got to be near the limit of what there is to know' is a gross misstatement. Our study of science is such an infant. Once we understand a niche of science, it gets developed to the extreme. The iPhone, computer technology, television, and all of the electronic sophistication of CT scans and MRIs lead us to believe we have conquered science while in reality we have barely scratched the surface.

Our view of human achievement reminds me of the view a dog once had of me. He was the smartest dog I have known. Would you like to know what he said one morning? [I can tell you suspect I am exaggerating. Well, I am; he was not an early riser. He did his profound talking in the afternoon.] Actually, what was interesting about this dog is he would get excited, in anticipation of what was about to happen, as I approached the tractor. He had high expectations. All day while I was away from home and the dog was around the house, the tractor was silent and motionless. When I returned and operated the tractor in the afternoon, it began blowing smoke, making noise, blinking lights, and even moving, pulling and pushing things. I was thinking, "What thoughts are going on in my dog's brilliant mind since the only difference between the tractor in action and the tractor in the dormant state of the earlier part of the day was me sitting on it?" WOW! He must think I am powerful! This is analogous to our view of humankind's current achievements. We are awed by the simple. I am not saying that I am not impressed by Hoover Dam, the Golden Gate Bridge, skyscrapers, the marvelous network of roads, internal combustion engines with associated engineering, and space-travel. We all should be impressed; however, the level of science that is at the root of all of the above is primitive.

We are still ignorant with respect to the big picture, i.e., gravity, quantum mechanics, space, time, motivation of electrons, creation, and the fundamentals of matter. The breakthroughs in science that bring change are given to prepared minds in 'AAHA!!' moments of inspiration. Such a moment will happen with respect to gravity when humankind is mature enough to cope, and the Anti-Gravity Revolution will begin. Wheeled vehicles will, like the horse and buggy, be relics of the past. Quantum mechanics suggest that if the location of a particle is unknown, it exists as a probability distribution and not an actual particle. This is the wave/particle duality. Einstein declared, "The more success the quantum theory has, the sillier it looks." Without Quantum Mechanics we would not have transistors, computers, iPhones, lasers, atomic clocks (which are accurate to 1 second in 20 million years) and more.

The conclusion of the book, that design influence is responsible for planet Earth and humankind, is based on statistical probability.

This book is written in the format of a Factoid, which is the miracle itself, followed by a Sermonette, which is a life lesson; it is dedicated to all fellow recipients of ordinary miracles. Read with an open mind and a thankful heart.

Family

Why is family important and what is its function? Several years ago in 1981, my wife (Ann), our two children (Larry was seven years old and Leigh Ann five), and I participated in a community luncheon at a neighbor's house when something life changing for me happened. We did not know it was a significant event until the following year.

After lunch the men gathered outside under a shady oak tree for conversation and for sharing our wisdom of how to solve the world's problems when a fifteen-year-old girl approached her dad and asked if she could go barefoot. He gave a short answer, "NO!" She was very displeased but said nothing. With a sad face she stomped off and rejoined the other community siblings. The following year, a new house trailer moved in down the road. It belonged to this young girl. She was pregnant, married, and a full member of the adult community. Last year, she could not make an independent decision of whether to wear shoes or not, and now she MUST make decisions which she has no experience base to draw from.

We noticed all teenagers enter a rebellious period when they begin the transition into adulthood. This allows them to demonstrate to themselves and others they are capable of making decisions. Usually the adolescent takes the diametrically opposed position of whomever the authority figure happens to be. "Son why don't you get a haircut?" is responded to, "I want my hair long!" Many issues are opposed by teenagers just to demonstrate this power of decision. After all, if they agree with the parent, they cannot be sure the decision is actually theirs.

Ann and I discussed how to avoid this in our children who were then six and eight years old. What we came up with was to quit child-rearing and begin adult-rearing. We decided the focus should be on the finished product and not the raw materials. If you have a broom factory, you would not refer to it as stick, straw, and wire plant. We had a family meeting where we explained that the children would never get another spanking from us. Their life is theirs to control and their decisions were final. We told them that we owed them whatever was required in time, money, and well-thought-out advice to ensure they grew to a self-supporting age and capability. We owed this because our parents had paid the bill for our dependent years and we would pay all we owed by providing for our children. It is a pay-forward system. You receive support from your parents or guardians then you in turn provide support to your children. Children owe nothing to the parents, but repay whatever they take as a child to their own children. We told them, how they turn out in life is their responsibility.

Can a parent really do this? We did. As odd as it sounds, in giving them more responsibility for themselves, I felt my responsibility for their development double. In order for them to make good decisions, they had to have information. How should I teach them not to play in the street? I took them to the street where a squirrel had been run over. I explained the squirrel had run in front of a car, and this was the result. I then requested if they should ever see a child that may be considering going into the street to warn them of the danger. I wanted them to feel as if they were custodians of keeping kids out of the road. Of all the jobs I have had, parenting has been the one with the most responsibility and the one I enjoyed most. I will not bore you with a volume of detail, but I will tell you of an interesting event.

One day Ann phoned me at work and told me Larry and Leigh Ann were having a horrific argument. I told her I could not do anything about it now, but without thought, I asked her to have the kids write down their respective issues and I would deal with them when I got home. Late that evening as I entered the house, they both approached me with their documents. What happened next was not planned. I simply said, "Swap notes." They did. Then I said, "Read." They did. Then they both looked at me as if I had a solution and unknown to me I did. I asked if they understood what the other said in the note. When they both said they did understand the notes, I told them to go into the living room and when they could come to agreement they could leave. It took them about thirty minutes of loud heated discussion, but they left each other that evening with a mutual respect that has endured. They were fifteen and thirteen years old at the time. I suppose I should tell you what the issues were. Leigh Ann asserted when Larry was around others he excluded her, and Larry maintained when she got into his Bronco she did not clean her shoes. The consensus was they did not treat each other with enough respect. They fixed it. I could not have taught them what they taught themselves. Adults must apply reason to their behavior, and this also applies to developing adults.

Adult-rearing was certainly not without issues, and I am a self-study student of the task, not a master. Our journey turned out well, and I thank God for helping with my deficiencies. We did diminish the rebellious years, and if I had a re-do on this one, I would start earlier. The end results are both are physicians and responsible contributors to society. I cannot say the adult rearing philosophy is cause–effect responsible, but it was a player in a game with a good outcome. To my wife, Ann, there are no shades of grey: all things are either black or white. Her considered opinion is I did not tell enough of the downside. After reading the section on family, the outside observer would think this philosophy is the utopic answer to family issues. She says based on her experience, she would not try to tell anyone how to parent. I love her, and she is a wise inside observer. It would require another book to give a blow-by-blow description of the past thirty-nine years, but the end results would be pleasing to any parent.

In the animal kingdom, several species form long-term family units to ensure that their offspring develop and survive. The Emperor Penguin is an example of a dedicated partnership to sustain the species. Their impressive reproductive commitment requires the male to incubate the egg for two months without eating after walking thirty to seventy-five miles inland in Antarctica where temperatures are minus forty degrees Fahrenheit with heavy winds. He keeps the egg on top of his feet, covered with his belly, while wobbling around in a huge circle with other males that are doing the same thing in order keep warmer while incubating their eggs. During this time the female has returned to sea to gain weight in order to have delicious regurgitated food ready for the chick when it hatches. This arrangement is difficult to imagine happening without design influence. Two birds that are ill equipped to walk, hike inland for up to seventy-five miles in the most inhospitable environment on Earth. They mate, and the female lays a single egg. Then how do they come up with a plan as complicated as, "Joe, you stay here and incubate our egg on top of your feet. I am going back to sea and will return in a couple of months when JUNIOR hatches. I love you too. Stay warm!"?

The family unit is necessary to ensure stability for adult-rearing. Having the ability to sustain and maintain a healthy nurturing family base is an ordinary miracle. We must consciously work to make the family unit functional. As Ann and I were maintaining the grounds of the church we attend (pictured on the book cover), I noticed several grave stones celebrated the life of the interred with the word MOTHER. My first thought was, 'What an honor.' Consider the wonder of motherhood. Your birth is one of your mother's contributions to the honor associated with the word MOTHER. Is your life one that would please your mother, filling her with pride of her contribution? If not, think about what you can do to make her addition to humankind meaningful with your remaining years. Brian Tracy in his book, *GETTING RICH YOUR OWN WAY*, asks, "What kind of world would my world be if everyone in it were just like me?" Make the person in the mirror someone that is respected by all.

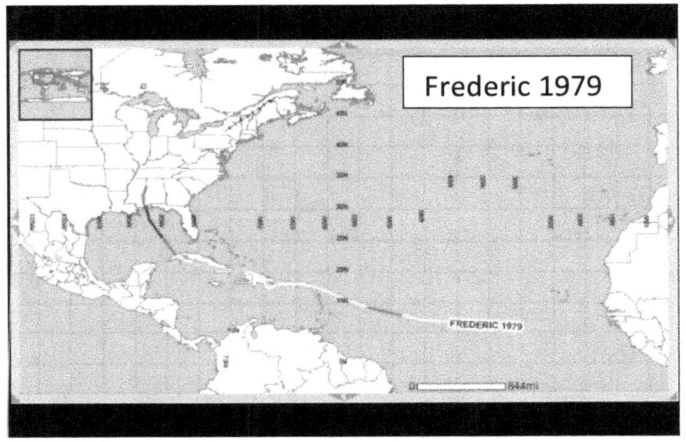

Hurricanes

Are hurricanes beneficial? When we look at a specific storm like 2005 Katrina where 1833 people died and damage was estimated at $108 billion, it is difficult to see a bright side. However, hurricanes play an important role in thermal equilibrium of the planet. The mechanism moves heat from the equatorial tropics northward. Many of the worst hurricanes that hit the United States are born off the west coast of Africa and travel west across the Atlantic drawing energy from the warm ocean water. The tranquility of the planet is sometimes interrupted by a climate-adjusting miracle that requires our acceptance. We are now capable of detecting and communicating an approaching hurricane with plenty of time for safe evacuation. Many coastal dwellers have grown to respect an impending storm and act accordingly while others tend to test fate and ride out the storm. I live approximately sixty miles north of the Gulf Coast and we seldom receive hurricane force winds. However, in 1979, Hurricane Frederic made landfall at Dauphin Island, Alabama and carried hurricane strength winds one hundred fifty (150) miles inland to near Meridian, Mississippi. This path brought the storm over our house. At first, my prayer was to protect my car, tractor, and boat from falling trees and flying debris. When the house shook, my attitude shifted to, "Take everything I own and leave this coffee table." Under there were my two children. This was a learning moment for me. I learned what the important things of life actually are. We came through the storm without significant damage. Five U.S. deaths resulted from Frederic, and four of these were inland. The only coastal fatality resulted from a person getting blown from a boat near Pensacola, Florida.

Standing Still

One of the most amazing miracles is the peace and tranquility of Earth's surface in the midst of all the fury of the universe. On a calm summer morning in south Mississippi, I sat in a boat on water that was as slick as glass, and not a breeze anywhere. Remembering that morning I thought, "This is impossible." With all the high-velocity movement that is occurring around us in the universe, it seems unlikely that our environment could be so calm. The earth is rotating, making one revolution every twenty-four hours. The distance around the earth at the equator is approximately 25,000 miles, which means if you are standing near the equator you are traveling over a thousand miles per hour. However, the sensation you would have is that of standing still. At the same time, the earth is 93 million miles from the sun and makes a complete orbit in 365 days. That orbital distance is 584 million miles, resulting in a velocity of 66,000 miles per hour. The sensation you have is that of standing still. The dynamics of the atmosphere of Earth rotating at the same velocity as the planet surface and space not having a drag taking away atmospheric gasses is nothing short of miraculous. Many other speed factors are also at play: the rotation of our galaxy, the expansion of the universe, and total movement of the universe. All of these factors have velocities that exceed 10^5 miles per hour, and there I sat in a small boat with no wind and no noticeable movement anywhere, with the exception of a fish striking occasionally and the birds playing on the nearby shore. If you do not find it miraculous that we are sheltered in the midst of all this motion, your belief in luck is strong (but will need to get even stronger before we are done).

In a sermon one Sunday, Pastor Cecil Locke told of an artist that was commissioned to draw a picture that depicted peace. The challenge was answered with several attempts that were magnificent drawings of peaceful, tranquil, pastoral settings. These included meadows with sheep, a shepherd, slow-flowing brooks, birds, flowers, and sunshine. Others included an autumn lake with magnificent reflections of the colorful vegetation along the shoreline, and a faint view of a lazy trout resting in shallow water. These were all rejected as not capturing the true essence of peace. Finally, the artist composed a scene of a violent storm with a lightning-loaded dark sky and huge waves with hurricane force winds from a heavy sea crashing onto the cliffs along the coast. In the midst of all this fury, there was a mother bird in a crevice, high on the shore wall, with her young under her wings in isolated peace. This not only won the artist his promised rewards for meeting the challenge, but fits well in depicting the protection we receive on the surface of planet Earth.

The way humans go at life is frequently referred to as the Rat Race. Wikipedia defines rat race as "an endless self-defeating or pointless pursuit. It conjures up the image of the futile efforts of a lab rat trying to escape while running around a maze or in a wheel." We could look at it as many rats in a maze all running in pursuit of a piece of cheese. In general, it is a great amount of activity with little or no accomplishment. Lily Tomlin said in the December 26, 1977 issue of *People* magazine, "The trouble with the rat race is that even if you win you're still a rat." The beauty of this world demands that we slow down and take time to smell the roses and observe the wonders of this great planet.

We spend many hours at work; overtime, stressful jobs, and commuting which leaves little time for family, friends, and enjoying the benefits of our economic gain. The main reason so many people participate in the 'race' is because we believe that it is necessary. Everyone feels the need to provide for their family, and this feeds the tendency to go full speed ahead without stopping to ensure we are on the right track. Most people we know are contributing to the rat race paradigm. Is it possible to succeed and not be a rat race contestant? If we are aware that we are a 'rat-racer,' then at least we can ratchet-back.

One thing we can do is evaluate how we are utilizing our time. Work, family, physical and spiritual health, and community are all worthy of consideration. What feels right for your life and goals? Another beneficial task is to schedule non-work related activities, just as we do the income-driven events. Schedule the time needed for your family. Give consideration to your spouse and other family members individually and collectively. Set aside time for self-reflection and exercise. I could never find time for daily exercise until I realized that thirty minutes before my usual time to get up and start each day, I had time. This became an enjoyable habit. Further application of time management will help escape the tempests of life. (We do know that it is not 'TIME' that we are managing, right?).

Ice Floats

Isn't it strange that ice floats? Not only is it strange, it is essential for our existence. Most everything contracts as it gets cold. Not water. Water is at its peak density at four degrees Centigrade, and this is critical to life on Earth. As water cools below four degrees Centigrade, it expands making ice less dense. If ice followed the rules for most everything else, it would become denser as the temperature dropped and sink to the bottom of the lake, pond, river, stream, brook, sea, or ocean. The consequences of this would be that the body of water would freeze solid during cold weather. When seasonal temperatures rose above freezing, thawing ice would form a water layer at the surface that would insulate the lower ice layer from warmth of the air and from the sun's radiation. Most water would remain frozen. This would result in no aquatic life and therefore no life on planet Earth.

When filling an iced tea glass in south Mississippi, it is common to say, "Just cover the ice." Of course, since ice floats, the glass is filled before the ice is covered. This happens in nature as well. When winter winds bring freezing weather, the cooling water remains at the surface, and as it freezes, it forms a barrier that protects the underlying body of water and aquatic life from freezing.

Many of our expectations in life are not fruitful because we are not looking closely at the details of design. We think ice is logically going to be denser than water and will sink, only to find ourselves rescued like a polar bear on a floating ice raft, in spite of our belief.

Which is heavier dry air or wet air?

The answer to this one is easy, i.e., ***clouds float*** in air. This is the miracle that makes rain possible. Air is made of 78% N_2, 21% O_2, and 1% Argon. Since water vapor (H_2O), with a molecular weight of 18, is much lighter than nitrogen (N_2), with a molecular weight of 28, clouds float. Wet air is much lighter than dry air. This is also why baseballs are hit farther in high humidity games.

One cubic yard of 85-degrees-Fahrenheit air can hold about an ounce of water at saturation. When air cools, the water vapor condenses, and we have rain. Approximately 3.1×10^{14} gallons of water is evaporated per day from all surface water, lakes, rivers, ponds, and mostly from the oceans into the atmosphere reservoir. The same amount is then sent back to the earth's surface in the form of fresh water through precipitation. This is a bubble of fresh water with a diameter of 8.6 miles delivered daily.

The measure of water vapor in air is humidity. Relative Humidity (RH) is the percent of saturation at any given temperature. As the temperature increases, the amount of water vapor that can be held in a given volume of air increases. The converse is also true: as the air temperature decreases, the amount of water it can hold decreases. If the RH is 100% (meaning the air is at the saturation point), and the temperature drops, we get precipitation, i.e., it rains.

This water cycle is important to life on Earth. Seventy (70) percent of the earth's surface is covered with water. If all water on, in, or above the earth were in a ball it would be 860 miles in diameter or 333 million cubic miles. About ninety-seven (97.5) percent of Earth's water is in saline lakes, saline ground water, or oceans. That leaves 2.5 % fresh water, and a majority of that (68%) is in glaciers and ice caps. That leaves less than one percent (0.8%) of Earth's water as "fresh" ground water or surface water. The recycled water that we depend on for life would make a ball 172 miles in diameter, and that is the amount that is globally available to use at any given time. The process that maintains the fresh water supply is the water cycle. Evaporation allows pure water to escape contaminated pools as water vapor then delivers it back to Earth as fresh water by condensation, i.e., rain or other forms of precipitation. If it were not for the water cycle, we would not exist.

Children enjoy hot sunny days and are disappointed with a cloudy, rainy sky. As we mature, we recognize that rain is needed to sustain life and the environment we enjoy. Renewal of fresh water is a good natural example of not becoming polluted with the debris of daily life. No matter how contaminated water becomes in the surface water reservoir, it is evaporated and cleaned in the recycle process. Do not let your life become and remain polluted. Water always has a chance to start over and so do you.

When is it too late to start over or start a new venture? The short answer is NEVER. Late is better than not at all. It is worth remembering that the future starts now. I forgot your birthday…. Do I make myself a note so I do not forget next year, or do I apologize and send a belated good thought? That was not really a question. Do it now.

Some late starters include this quote from Wikipedia: "Sylvester Stallone was 30 when he wrote & starred in the first *Rocky*. All throughout his life, he has pushed his body through rigorous training routines for his film roles. Most notably, at age 43 he developed his now famous *Rambo 3* physique which got him named as 'body of the 80's.' Now in his mid-60's, he's still pushing himself physically with his muscular physique which can be seen in the *Expendables*."

Colonel Sanders started the *Kentucky Fried Chicken* franchise when he was in his sixties. The first KFC franchise opened in Utah in 1952, and grew rapidly expanding across the United States and abroad. He sold the company in 1964 for $2 million.

This section brought on a poetic moment:

TODAY IS THE MOMENT WHERE THE FUTURE BEGINS

Yesterday was once a tomorrow,
And it now holds our memories,
Like it once held our dreams.

Today is just a moment where the future begins,
And tomorrow we view whatever we send.
You are older today than you have ever been,
 And still you are younger than you will ever be again.

So take advantage of your youth, regardless of your years;
Today the window of opportunity is here.
Starting now may put you clearly at the rear;
However, you're ahead of those whose start is still near.

Yesterday is gone, and tomorrow never comes;
When action is required, today is the one.
Yesterday is nice when looking where we've been,
Tomorrow is good when planning where to end,
But today is the moment where the future begins.

Snowflakes

What a gentle way to get frozen water to the ground! The standard method could have been baseball size hail falling from two miles up. Ice blocks called megacryometeors have fallen from the sky weighing over 100 lbs. Besides falling gently, snow also has other beneficial attributes. The primary one is insulation: protecting the ground from harsher freezes and preventing cycles of thawing and refreezing that could damage root systems. Why would water not condense into frozen droplets instead of millions of similar but unique snowflakes?

Does this qualify as a coincidental occurrence or is a design element apparent? Rain droplets freezing would appear to be a more likely random chance scenario and if this were the case, the results would be devastating. Ice bullets would be the cold weather perception. When large hail does occur, it makes the news. It is rare and life threatening. The miracle of snowflakes is a wonderful gift.

A child asked, "Where does all of the white go when snow melts?" It is curious why water in liquid form is translucent, thus clear, and in the form of snow it is white. An explanation requires an understanding that the color an object appears is white minus the frequencies that are absorbed. Since the electrons of the molecules of the object being viewed are excited by some specific frequencies within the visible spectrum, these are absorbed. The other wavelengths are reflected, and those reflected wavelengths are responsible for the color we perceive. Snow is composed of many small translucent ice crystals, each bending the path of light as it allows it to pass through. The bending occurs until eventually the light exits the cluster with no specific frequencies being absorbed.

Cycles

Have you ever noticed how many new beginnings we see in nature? Each morning brings a new day, every 30.4 days on average we see a new month, every 365 days starts a new Spring, Summer, Fall, and Winter. Every rain brings refreshing new, clean water. Plants and animals die and recycled into new life. A baby is born as an ancestor dies. Plants bloom many times through their life; solar systems recycle; and atmospheric nitrogen and oxygen cycle. You can think of others because they are plentiful and an essential part of design.

We cycle things in our life. A car needs new tires, and we trade it in on a new model. We start over several times in our business careers. The average tenure in jobs is about four years in the United States. Sometimes starting over is a good thing to do, i.e., a chance to begin fresh and do it better this time. But, do we sometimes recycle too much? PolitiFact.com estimated in 2012, the lifelong probability of a marriage ending in divorce is 40%–50%. The family unit needs to be a stable institution to provide a society-building foundation.

We also see cycles in society that tend to be generation driven. My mother remembered as a child having to wait until the adults had eaten to get lunch at the children's table. The only parts of the chicken that remained were the back, neck, and wings. As she grew to adulthood and had children of her own, she decided that they would not have the same fate. Her children got the breast, thighs, and drumsticks while she continued to eat the back, neck, and wings. In turn, her children always enjoyed the better parts of the chicken and placed no significance on the cut of meat that their children received; therefore, their children returned to their grandmother's plight of backs, necks, and wings. This section is not about how we eat chicken, but rather about why things cycle from generation to generation. When we take for granted that which our forefathers fought to preserve, we walk a slippery slope.

The Serenity Prayer is the common name for an originally untitled prayer by the American theologian Reinhold Niebuhr (1892–1971). It has become part of the twelve-step program of Alcoholics Anonymous. The most common version is, "God, grant me the <u>serenity</u> to accept the things I cannot (should not) change, <u>courage</u> to change the things I can (should), and <u>wisdom</u> to know the difference." 'Should' and 'should not' are alternative readings to simply say that we should not change a good thing to bad just because we can.

Good and bad are relative terms that vary greatly among groups. A concept of good that I enjoy is, **'That which is conducive to the survival of humankind is good, and conversely, anything that is detrimental to the survival of humankind is bad.'** Since good and bad are relative terms, the less corruption that is in your core principles, the better your 'good' will be. The vices you support in moderation will appear in your children in excess.

Accepting things that we should not or cannot change will make life less worrisome, and worry is a nonproductive waste of time. Worry is negative and should be replaced with positive thoughts. What can I do to improve the outcome? The Chinese proverb, **"It is better to light one small candle than to curse the darkness"** is worthy of application to anything you worry about. Mild worry can have positive effects if it heightens awareness and causes preventive measures to prevent the envisioned danger. If you worry about being attacked by a dangerous animal and it prompts you to not anger the large Rottweiler next door, that could be a good thing. Worry that does not prompt positive action is a waste of good time and mental power. It is thinking about an outcome that you do not want to happen. It is far better to replace that with a thought of an outcome that you would prefer.

Humankind has progressed to recognizing recycling as a good objective. We now recycle paper, plastic, glass, and metal, turning waste into new products. One of the hindrances to recycling is products are not manufactured with recycling in mind. A potential solution of sustainable design is presented in the book, *Cradle to Cradle: Remaking the Way We Make Things* by architect William McDonough and chemist Michael Braungart. They suggest every product (and all packaging they require) should have a complete "closed-loop" cycle mapped out for each component, i.e., a way in which every component will either return to the natural ecosystem through biodegradation or be recycled indefinitely.

It is pointless to look back and question decisions of the past, wishing we had taken alternate career paths or wishing we had made different social decisions. The ability to entertain such a thought means we are currently alive. The decisions we have made have produced a path that has allowed us to arrive at this point in time. We have no guarantee we would still be alive if different paths were chosen in our past. I have a philosophy: **If you are alive, your past has served you well.** The awareness that things could have been done better is just a beacon to start now and make the future your dream. Since you cannot change the past, learn from it and go forward.

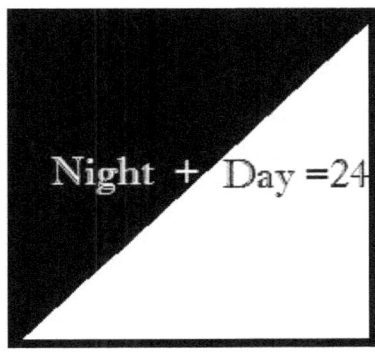

Day and Night

The earth makes one rotation on its axis every 24 hours. This is a miracle of scale. The half of the planet that is facing the sun experiences day, and the half that is away from the sun is in darkness or night. Larry Jr. once asked his grandfather, "How many hours are in a day?" When he received the answer of 24 he then asked, "How many hours are in a night?" That aside, what if the earth spun as slow as Venus, once every 225 earth-days which results in 1.96 days per year? Or, we could have the other extreme of a 10 hours rotation like Jupiter where night and day are each five hours long. Mercury has one day every two years. Life would be very different, maybe too different for us to exist.

A 24-hour day works well giving time for work, play, and rest. It gives a natural clock for starting activities and for resting. This is known as the circadian clock. There are many biological circadian rhythms that exist in plants, animals, fungi, and cyanobacteria. Circadian rhythms are endogenous ("built-in", self-sustained). They are adjusted (entrained) to the local environment by external cues called zeitgebers, commonly the most important of which is daylight. This information is from a good read in Wikipedia. Of course, the body would evolve to take advantage of the repeating periods of daylight and darkness, but the twelve hours of each is a nice touch. This is one more clue that we live on the Goldilocks planet, i.e., 'Everything is just right.' Can you see design influence yet, or do you still think we are just lucky?

Night is the period of time the earth is in its own shadow. It is interesting that darkness is not an entity; it is rather the absence of something: light. Human life goes in cycles. We have periods of growth where we have abundant energy and strength to accomplish our most difficult goals. This is followed by lower, more dormant periods where we are in the shadow of our last accomplishment or at a plateau of a still to-be-reached goal. These are not times of weakness but rather a time to rekindle and rethink our path for the next moment of light.

Nitrogen Atmosphere

Nitrogen, symbol **N**, is the chemical element of atomic number seven. As the major component of Earth's atmosphere, it is a colorless and odorless gas in the form of N_2. Nitrogen is a common element in the universe, estimated at about seventh in total abundance in our galaxy and Solar System. The earth's atmosphere is a thin layer of gas that surrounds the earth supplying breathable air as well as insulation from extreme temperatures and providing protection from ultraviolent radiation. From Wikipedia, the five layers of Earth's atmosphere from lowest to highest are: Troposphere (0-7 miles), Stratosphere (7-31 miles), Mesosphere (31-50 miles), Thermosphere (50-440 miles), and Exosphere (440-6,200 miles). The atmosphere becomes thinner with increasing altitude, with no definite boundary between the atmosphere and outer space. This life-giving bubble is about 300 miles thick, through the Thermosphere, and comprises approximately 78% Nitrogen, 21% Oxygen, 0.9% Argon, and 0.03% Carbon Dioxide. The Troposphere is about 7 miles thick and contains 80% of the mass of the atmosphere. At sea level, the pressure is 14.7 psi. The pressure and amount of gas decrease as you extend outward from the earth. How much of a coincidence is it that the volume of nitrogen is the amount needed to give thermal insulation and provide a vehicle for the water cycle on Earth? Another stretch, of equal magnitude, for coincidence is that the weight of N_2 (28) is heavier than H_2O vapor (18) which allows clouds to float and provides fresh water for life on the surface. The third most abundant gas in the atmosphere is Argon which has a molecular weight of 18. If Argon was the predominate gas of the atmosphere, water vapor would not float, thus there would be no rain cycle. Is this a coincidence or an ordinary miracle?

Nitrogen has both the characteristics and the volume to make the atmosphere of Earth functional in order to support life. How are your characteristics and their volume balanced in your life? John Denver sang, "... I fiddle when I can, work when I should." Consider the characteristics and their volume in your life. Create a balance that will serve you and those in your sphere of influence.

Nitrogen Cycle

Nitrogen is 78% of the earth's atmosphere and is continuously being depleted and replenished. Nitrogen's presence in the atmosphere is in the form of a very stable molecule of N_2. This is the result of a triple bond common to nitrogen with five electrons in its valence shell. Not being reactive is great for a stable atmosphere; however, it also places a significant challenge on all life that depends on nitrogen compounds for survival. Nitrogen is a critical part of amino acids. The conversion of atmospheric N_2 into a form useable by plants and animals is called fixation. This process is primarily carried out by bacteria such as those in the roots of nitrogen fixing legumes and to some extent by lightning. A stable atmosphere of nitrogen requires that it be nonreactive, while at the same time, for living organisms, it must be very reactive. To have an atmosphere of nonreactive nitrogen and also have it as a very reactive component of living organisms is an ordinary miracle.

We also need to be nonreactive, i.e., tolerant of others who are different from us. Recognize that children are influenced and shaped by their circumstances, parents/guardians, peer group, role models, teachers/education, faith/belief-system, and socioeconomic status. As we enter adulthood these influences continue as factors; however, our developing spirit allows paradigm shifts, sometimes from a single ray of light or exposure to a specific catalyst/enzyme. We see a better way and plot a course of change. A nonreactive atmosphere of tolerance, from a stable society, can assist the human fixation process. Remember, nitrogen fixation is the process by which nitrogen is converted from its inert molecular form to a compound more readily available and useful to living organisms.

Energy Availability

Is it strange that energy is stored in such convenient packages that are easily accessible to man, i.e., wood, oil, and coal? Why wouldn't natural evolution recycle available energy to support emerging life quickly? Instead, the kinetic energy of the sun for thousands of years is absorbed by plant life and converted into potential energy that we tap today. I have a wood heater, and it occurred to me that we are warmed by the sun's energy that was stored in a tree over several decades. Man is the only animal that is so energy dependent. We are the fastest, strongest, smartest critters on Earth largely resulting from our use of stored energy. No animal can out run, out swim, out fly, overpower, or out think us because we compensate our natural abilities with creative energy using solutions.

Collectively we have been able to utilize all of the mentioned energy stores plus hydroelectric, wind, chemical, solar, and nuclear. This utilization reduces manual labor, increases mobility, and allows humankind to be more specialized. My small two-row farm tractor has forty horsepower. I do not do the work of forty horses; however, the reduction in manual labor is obvious. Multiply this by several billion, and you have the global impact of our energy utilization.

The Anomaly of Oxygen (availability as O_2)

Oxygen is the second most electronegative element second only to Fluorine. It is, therefore, very reactive and forms oxides with almost all other elements. This means that oxygen as we breathe it, O_2, cannot exist in the atmosphere especially at the 21% level except for the miracle of photosynthesis. The miracle that is manifested in the cabbage leaf picture is liberating O_2 from CO_2 ubiquitously over the planet. One calculation maintained that an acre of trees would produce enough oxygen to sustain about 18 people.

Plant life is a reduction process that converts carbon dioxide, CO_2, into cellulose, sugar, and O_2. Who thought of that little twist in the great scheme of things? Plants use CO_2 as it becomes available and maintains the atmospheric concentration at approximately 400 parts per million. The amount of CO_2 that is annually converted into biomass is roughly 100×10^{12} kg, which results in 18×10^{12} kg O_2. It is interesting that the amount of fossil fuel that was used globally in 2009 was 4×10^{12} kg. That is 4% of the annual amount of biomass. I have a suspicion that there is equilibrium between crude oil and the production of biomass that is facilitated by deep ocean currents and high pressure that converts annual biomass into an equal amount of crude oil. That would make crude oil a renewable resource. The deepest part of the oceans is the Mariana Trench near Guam with a depth of 11,000 meters (6.8 miles) where pressure is greater than 16 thousand psi. With thermal vents, parts of the deep sea are 300 degrees Fahrenheit. If you are an engineer looking for a place to manufacture crude oil, this is a candidate. Add to these deep-sea ocean currents, and one has a vehicle for getting the raw material of biomass to the reaction chamber. Does this explain why crude oil is stored deep, beneath the bedrock, instead of floating on the surface as it would if it were generated near the surface by decaying surface life? Fossil fuel is generated in an anaerobic (absence of oxygen) process. The deep ocean environment is perfect.

Balance is such an essential part of our existence. As we deplete O_2 from the atmosphere, it is replaced by photosynthesis. Our lives need balance. I have a tendency to be obsessive compulsive. When I start a project, I lose the balance needed to be the all-around flexible family man. We all need to address the issues that disturb the balance in our lives. Do we eat to live or live to eat? Do we exercise routinely to maintain physical fitness? All work and no play make Jack a dull lad; however, all play and no work leaves Jack unpaid and sad. I spoke of balance earlier. It is reoccurring in nature and applies to many systems. We need balance in many areas of life.

The Anomaly of Oxygen (abundance)

Oxygen is the third most abundant element in the universe following Hydrogen and Helium. Doesn't it seem strange that the element with the atomic number of eight is the third most abundant? Hydrogen, Helium, Lithium, Beryllium, Boron, Carbon, Nitrogen, and Oxygen are the first eight elements on the periodic chart in order from lightest to heaviest. One would expect the nuclear furnace of stars would combine Hydrogen/Hydrogen to make Helium then Hydrogen/Helium to make Lithium and progress up the atomic number scale instead of going one then two then eight. We need oxygen for life. How do you suppose that happened?

Are there things in your life that are very pleasing but unexpected? You just happened to be in the right place at the right time.

Since this photograph was taken on Earth, this is probably water!

Pools of Water

Compounds exist as solid, liquid, or gas depending on temperature and pressure. Isn't it strange that the only liquid you come upon as a pool, pond, puddle, lake, river, stream, creek, brook, ocean, or sea is water? How convenient that this life supporting liquid is so unique! No other liquid competes with water. Why wouldn't some other compound be abundant in liquid phase at Earth's surface? This has to be design. Other liquids that we use routinely (including petroleum crude) are less dense than water and logically should be floating on all surface water; however, these are locked deep beneath the surface. Flip Wilson always said, "What you see is what you get." I look at water and see a miracle. What do you see?

Water is unique and so are you. What are your strengths? Find them and work to make them better. I attended Ed Foreman's 3-day Successful Life course and one of the philosophies he teaches for self-improvement is developing good habits. His tool for doing this is to force yourself to practice correctly whatever it is that you would like to improve for 30 days and it becomes a habit. It gets easier as the days click off, and by day 30 you are no longer forcing the behavior, it is a habit. Try it. It works for anything that you want to improve from eating better, being neater, improving organization, smiling, being friendly, being punctual, and whatever else you want to improve.

Colonial Williamsburg

The Well

When thinking of water to drink, we think of a well before we think of a spring, creek, river, or lake. Or to say another way, we think of ground water before surface water. The reason is obvious: surface water gets contaminated. Garrison Keillor usually ended his *A Prairie Home Companion* broadcast with a quote from Will Rogers, "Always drink upstream from the herd."

That brings us to a God-given friend, i.e., the aquifer. An aquifer is porous rock and sand that hold water and allows it to flow free from surface contamination. These aquifers are present almost anywhere on Earth at some depth. Is it just a coincidence that the earth has a protective crust over an aquifer? From biblical times and continuing today, we use wells to tap the aquifers for life-giving water. Ninety-five percent of rural USA depends on this water as a source for drinking water. My children and I were raised on it, and I appreciate this gift as an ordinary miracle.

Some of the good things of life require an effort on our part. Digging is required to get to aquifer water, and digging is also required to get best use of your talents and natural gifts.

Running Water Uphill

Everyone knows water runs downhill, or as my geologist friend said, *"Water moves down the hydraulic gradient. Pressurized water, such as in an artesian aquifer, can go uphill."* This in layman's terms: 'Water seeks its own level.' Gravity is attracting, or pulling, or pushing toward the center of Earth. So how does water get from under the ground to the top of a tree since it is not due to water moving down the hydraulic gradient? The miraculous answer is capillary action. We explain it with forces of adhesion, cohesion, and surface tension. Water has an affinity for the inside materials that make up a plant root. This adhesion allows water to start its climb up the plant. Water molecules also have cohesion for other water molecules, pulling others along the journey. It is pretty amazing to witness the ordinary miracle of water flowing uphill to the top of a 90-foot-tall longleaf yellow pine. A pump that pulls a perfect vacuum can only lift water to a maximum of 33.9 feet. A 90-foot-tall pine tree uses a ballpark estimate of 50 gallons/day. This rate represents a substantial uphill flow.

It is difficult to have it both ways. Water flows downhill. Under the influence of gravity, it moves closer to the center of the earth. Then all of a sudden, the need is to flow against gravity and climb a tree. This is a unique property of water's hydrogen bonding that gives it exceptional adhesion characteristics. Could it be that such a pivotal property is not part of design?

We have a tendency to remain in our comfort zone because it is easier and safer; however, if we are to grow, we must advance against the tide and like water, figuratively speaking, dare to climb a tall tree.

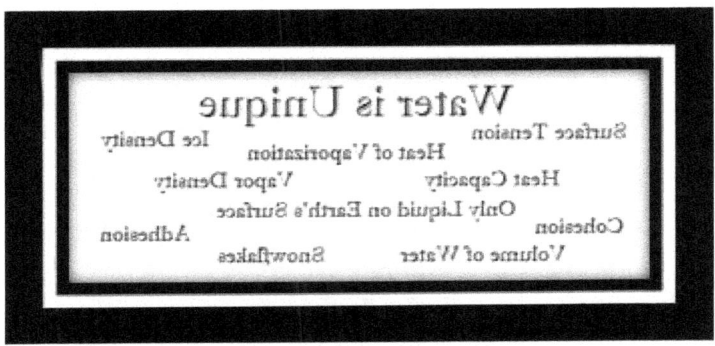

(Water is unique and difficult to read by the casual observer)

Water Anomalies

1. Practically the only liquid you will find on Earth's surface is water. (Essential for life on Earth)
2. When water freezes, it expands. (Essential for life on Earth)
3. Snowflakes float to Earth instead of ice bullets shooting to the ground.
4. Water's vapor phase is lighter than air, enabling the water cycle, i.e., rain. (Essential for life on Earth)
5. The cohesion, adhesion, and surface tension properties of water are unique. This allows water to run uphill, i.e., by capillary action water climbs to the top of the tallest tree. (Essential for life on Earth)
6. The heat capacity of water is higher than most anything, absorbing a great amount of energy with little change in temperature which regulates

temperature during night to day cycles. (Essential for life on Earth)
7. The abundance of water on Earth is ~333 million cubic miles or just enough to sustain life.
8. Fresh water is available in aquifers almost anywhere on Earth. (Essential for life on Earth)
9. High heat of evaporation facilitates the earth's heat equilibrium and distribution.
10. Water covers 70% of the earth's surface, a great water/land ratio.

Why is water, an essential compound for human life, so different from other compounds, and why are these differences clustered around needs to support life? Why is the amount of water on Earth just the right amount for our existence? Water is suspiciously different enough from other compounds to suggest design influence. As a gas, water is one of the lightest known; as a liquid, it is much denser than expected; and as a solid, it is much lighter than anything when compared respectively to the liquid form. These anomalies give rise to a continuous fresh water supply, prevent frozen water bodies, and stabilize the earth's temperature. The heat capacity of water is huge, beyond expected. To raise the temperature of one gram of water one degree Centigrade (by definition) requires one calorie of heat, which is more than ten times the amount of heat that is required to raise a gram of copper one degree Centigrade. Water can absorb a great amount of energy without changing temperature very much. This high heat capacity helps stabilize our body temperature. Water also has an unusually high heat of evaporation. It requires five times the amount of energy to vaporize a given amount of water as it does to raise that amount from zero to 100 degrees Centigrade. Another anomaly of water is the aquifer where water is protected from surface contamination is tucked away, mostly available to man alone. Water cohesion, adhesion, and surface tension are unique and allow water to climb a tree through capillary action. Water forms snowflakes to get frozen water from the atmosphere to Earth's surface in a gentle manner.

This section is somewhat repetitious, but necessary, in order to present an encompassing view of the miraculous attributes of water. All of the properties of water are extraordinary and all of the distinct characteristics are essential for life as we know it. Water is a super hero. How does water compare to Nitrogen, the Iron Core, Ozone Layer, or the Tilted Axis in importance to life on Earth? It should be easy to get agreement that this is a silly question. If they are all essential, then they are all critically important. They cannot be compared. They are all necessary.

When looking at the human population, we do not always apply the same logic. We tend to place value to different individuals. Some people are worth more in the business environment due to their knowledge and/or their skill set. The CEO is paid more than the entry level production worker. Both are essential, and they are compensated according to their contribution to the whole. However, their respective values as individuals are the same.

Consider this example: You have a $60,000 automobile that will not run, because it has no spark plugs which you can purchase for $15.95. The plugs are worth around sixteen dollars; however, without them, the car is useless as a motor vehicle. It could still be used as cramped living quarters, but it is not functional as a vehicle. The spark plugs have the same value to the car as any of the other essential and more expensive parts, such as the $800 ignition module.

I have a thought when looking at people. We all add up to 100. Where one has an extremely high IQ, another has superior physical strength. All people possess different strengths and weaknesses in cognitive, physical, and social arenas. That does not make anyone better than another – it just makes everyone unique.

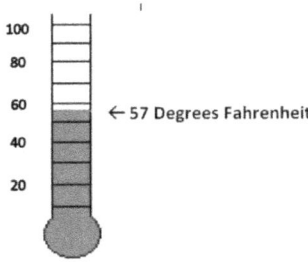

Mean Global Surface Temperature

The combined land and ocean global annual temperature for 2011 was 57 degrees Fahrenheit. Since water freezes at 32 and boils at 212 degrees Fahrenheit, 57 degrees is a comfortable mean temperature to maintaining life-giving water in the liquid phase. The delicate balance to keep the earth's orbital distance and the sun's energy output at the needed values to achieve this is beyond chance. I am thankful for the miracle.

Earth has been called the Goldilocks planet: 'Everything is perfect: not too hot, not too cool, but just right.' CO_2 is another player that contributes to the mean global temperature, and is considered to be the thermostat of the atmosphere. It is known as the major greenhouse gas (GHG). The other GHGs include water vapor, methane, nitrous oxide, and ozone. The GHGs absorb and radiate energy in the thermal infrared region. Without this regulating effect, the mean global surface temperature would be near 33 degrees Centigrade cooler and drop below zero degrees Fahrenheit. Antarctica is another regulator due to the large amount of sun radiation that is reflected by the white surface.

What are the regulators in your life? Look for them, develop them to a good regulating level, and make a habit of using them. Keeping your body mass under control, being careful of what you breathe, and tweaking the speed you get around will also help.

Ecosystem Balance

In order to have a sustained ecosystem there must be balance. We have looked at some of the components of our ecosystem system that are consumed and continuously replenished. These include O_2, N_2, fresh water, surface heat, and the ozone layer. There are many more components of our environment that require a state of equilibrium; however, the beauty of the *Ordinary Miracle* of balance can be demonstrated with these.

It is not enough to simply have a system replenish the components that are required for life, but the rate of replacement must equal the rate of consumption. When I was in grade school, we were taught the components of Earth's atmosphere were 78% nitrogen, 21% oxygen, and 1% argon. This balance has existed for approximately 200 million years. There is equilibrium where the rates of depletion are met with restoration. The current debate of greenhouse gas concentration increase is interesting. The concentration of CO_2 in the atmosphere is important due to the impact it has on the mean global surface temperature. The concentration has risen due to humans using the heat of combustion to provide energy to drive almost everything. If it will burn, we burn it releasing the long term stored CO_2 from oil, gas, coal, and wood back into the atmosphere. By doing this have we disturbed the balance of the ecosystem?

Is the planet temperature going to increase and melt the polar ice caps? If the mile-thick layer of ice that covers Antarctica melted, the oceans would rise an estimated 200 feet. Conversely, the North Pole ice floats on the Arctic Ocean and melting would not affect ocean levels. Melting of the polar ice would not be a good thing and is not on the horizon as a pending doom. The reason CO_2 is measured in parts per million (ppm) is because plant life is consuming and converting it into complex carbohydrates as it becomes available. The past 100 years have certainly given a bump in CO_2 release. From studies using trapped air bubbles in glacier ice, scientist are concerned because CO_2 levels are the highest in the past 400,000 years; however, plant life is a busy regulator and will reach new equilibrium before the balance is significantly disturbed. Higher CO_2 levels are a welcomed food source aggressively sought by all plant life with voracious appetites.

I do not advocate abandonment of being a good steward in care of our planet. We were given a great gift, and with every privilege there is accompanying responsibility. I challenge you to appreciate balance that is built into Earth and do your part to live in harmony within our ecosystem. Small steps by each individual makes big strides collectively.

Is it coincidental that atmospheric concentrations have reached equilibrium at ratios that are ideal for life? Any environment that supported life evolution would be ideal for that life; however, Earth's environment with free oxygen stabilized at 21% is ideal for biological development as well as combustion support.

Not only is the ratio ideal and stable but the volumes are stable over time. The ratios could have equilibrium independent of volume if the separate cycles were not in balance. This is important in order to support the inherent functional essentials of the atmosphere, i.e., water cycle, ozone layer, and thermal regulator. The atmosphere is a collective miracle that is filled with independent miracles.

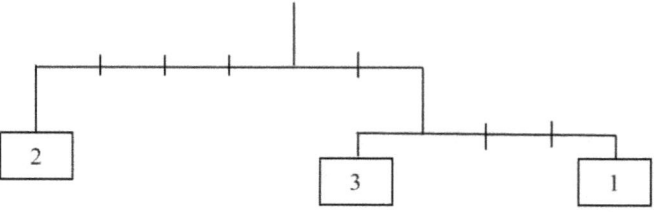

The above puzzle is balanced: 2 x 4 = 2 x (3+1)

A look at Erich Friedman's balance puzzles in Wikipedia 'Balance' section, similar to the one above, is an interesting mental exercise. The complexity of balance can easily be seen as the variables increase. The solution to balance in our ecosystem is an exponentially higher order of magnitude than any of these.

A balanced life can be a challenge as well. We would be well served to observe the weekly day of rest. The more complex our life, the more complicated the balance challenge. Just as in the puzzle above, more variables increase the difficulty in achieving a balanced life.

Gravity in action

The Miracle of Gravity

One of the most extraordinary phenomena of physics is gravity, a miracle that we do not begin to understand. It holds the planets and stars together; it holds things on the ground; it holds the moon in orbit of the earth; it holds the earth in orbit of the sun. The amazing thing about gravity is we know what it does and the laws that describe its effect. What we do not know is what gravity is. Is it a wave force? Einstein spent the last thirty years of his life trying to figure it out. With all of the tools that scientists now have available, we still do not know how to influence gravity. If we did, we could shield ½ of a wheel from gravity and have a motor. Gravity is a mystery force. I think I know why we do not yet understand it and talked about that in the Inspiration section.

What holds your feet to the ground? What gives you the traction that you need to work and contribute to the good of the whole? We all need encouragement to help steer in a healthy, fruitful direction, and we need to share guiding forces with those within our sphere of influence. Why have many lost their traction and instead of forging forward for achievements are sitting back waiting for a handout?

Inspiration

The reason no one knows or understands gravity is that we as humankind are not ready to deal with the knowledge. When we are, someone will have an 'AAHA!! Moment' and will be gifted with the inspiration to move us into the next technological revolution. If you could block gravity to half a wheel, you have a motor. Block gravity to a rug, and you have a flying carpet. One day, wheeled vehicles will be a thing of the past. I have grandchildren that are capable of such an AAHA!! Moment, but if not one of them: someone, someday.

Look around your world at common things that are extraordinary new comers. Electricity is the first in my mind. We use electricity for lighting, heating, transportation, communication, and computation. People were familiar with the shock given by several species of fish as early as 600 BC, but the understanding of electricity did not occur until the early nineteenth century. Still, the use of electricity did not start until the late nineteenth century. Several years ago in the wee hours of morning while working night shift at a rayon manufacturing plant, I asked the shift manager, *"If you were left alone naked in the forest, how long would it take you to come riding out in an automobile wearing a wrist watch?"* He replied, *"I would never in a long lifetime."* The ability, intellect, and inspiration to progress from awareness, understanding, and utilization to dependency is miraculous. What is the source of the inspiration and the vision of the potential? It is reported that twenty-two people before Thomas Edison had made a light bulb; however, it did not become an available form of lighting until he developed an integrated system of lights, generator, and distribution main. We are a gifted creature making quantum leaps by stumbling along an inspired path. For man to build what has been built is miraculous. Can you imagine standing on the Colorado River bank at Black Canyon and visualizing the 1200-foot-wide gorge filled with the 700-foot-tall Hoover Dam? It is 660 feet wide at the base with a 45-foot-wide highway at the top. Lake Mead is the result, which is 500 feet deep at the dam and provides hydroelectric water to generate 1.3 billion watts of electricity. President Hoover's name was given to the dam, but Frank Crowe was the chief engineer with the vision to oversee the successful project. Guess where inspiration originates? When we are ready for anti-gravity the 'AAHA!!' will be given.

Ten eureka moments are cited at (http://science.discovery.com/famous-scientists-discoveries/10-eureka-moments.htm). They include Einstein and Relativity, Tesla and Alternating Current, Archimedes determining the density of gold by water displacement, and seven others.

People are often inspired by 'Mother Nature.' Isn't it strange that we personify nature? Could it be because nature actually communicates with us? In the movie *Avatar,* the native inhabitants of the planet Pandora, had a physical connection to other animals and also to plant life through receptors in their hair. As I watched the movie, I was not surprised by this communication because I've often felt a communication with nature. All life is connected, and I believe the reason for the connection is we are all part of a universal awareness that will be talked about in a later section.

Subliminal communication is on display in a meeting or assembly where the atmosphere seemed to be electrified. It seemed to be easy to communicate. Everyone is on the same page and tuned-in as if they have a single mind? People are seldom aware of this subliminal communication. Most often we just sense a friendly or hostile connection with another person or a group.

I think subliminal communication is more prevalent in lower animals than in humans. We have made such good use of verbal communication that we have, to an extent, suppressed subliminal abilities. I have noticed that when two dogs approach each other, they can sense hostility almost as far as they can see the other animal. Sometimes in meeting another dog, bristles will raise and they will take on an aggressive posture; at other times they will wag their tail in a friendly manner. Is this subliminal communication, or are they sending physical gestures?

The story of the hundredth monkey effect was published in Lawrence Blair's *Rhythms of Vision* (1975). The claim is that scientists were conducting a study of macaque monkeys on the Japanese island of Koshima in 1952. The monkeys were fed sweet potatoes that were dumped on the sandy soil. One female monkey washed her potatoes before eating, and this skill was then learned by her young. Others learned from observing, and eventually it became common practice. Once many monkeys, 'the-hundredth-monkey,' had the potato-washing knowledge, monkeys on neighboring islands became aware and started washing. Those skeptical of subliminal communication suggest that a potato-washing monkey swam to a neighboring island, and they picked up the habit by observation.

Have you ever had an inspiration to do something? This book is the product of inspiration. Much of my inspiration is derived from observing how our lives are dependent on and interacts with nature. To communicate *Ordinary Miracles* was an AAHA!! Moment.

The Miracle of Words

We overlook the miracle of words. The power of words is referenced in John 1:1 "**In the beginning was the Word**, and **the Word was** with God, and **the Word was** God." Is this a real miracle or just a random occurrence of evolution? Many animals have a brain similar to the one you have. But no animal thinks like man. The difference is we have words. We think in words; we communicate in words; we record concepts and past events in words. Isaac was tricked into giving a blessing to Jacob he had intended for Esau, but he could not change it after discovering the error because he had spoken the words. Do words have power beyond the influence levied on the person that hears them? Esau thought so.

It was not so long ago, feminists became vocal about using generic gender for professions instead of being male specific. Policemen become 'the police'; postman, 'mail carrier'; chairman, 'the chair'; fireman, 'fire fighter'; mankind, 'humankind'; and more. Gradually the norm changed to the more generic references. Did this impact the hiring of more females into these functions or did the trend of hiring cause acceptance of the new words? The words came first then the hiring. According to Terese M. Floren in her article "History of Women in Firefighting" (2007), two women marked the entry of paid professional females in the firefighting community in 1973-74, and today there are more than 6,500 paid female firefighters and fire officers.

I know people who came to this country and say they knew they were American, after a period of time, when they began to think in English. Language is the vehicle for thought. The language we speak is not as significant to our thought process as words. It does not matter if the words are English, German, Russian, French, Spanish, etc. or of a specific dialect within a language. If your thought can be represented linguistically, you are empowered.

Children are sometimes cruel when speaking to, or about, their peers. Ethnic or physical attribute slurs are common. A child lacks experience to realize the impact of their words. The sad part of this argument is they obtain the cruel speech by listening and observing adults. It may be difficult to eliminate the impact of these words, but perhaps we could all reach a level that a crazy label will not affect us. I like myself, so any word that references me is a good word. Whatever our heritage, or current condition, happens to be, it has successfully woven a path to our existence. We can reflect on both our personal and collective history with a thankful heart because the end result has produced each of us. We have opportunity because of, and in spite of, our heritage. Now, the challenge is for each to live in such a way that future generations will view their heritage with pride.

Speak your words with care. Your mother told you, "If you can't say something good, don't say anything at all." Do not speak your bad thoughts. Words have power.

My four-year-old granddaughter, Jordan, placed a headband on me and left it there for over an hour, walking by every ten minutes or so to inspect and laugh. I finally removed it and laid it by my chair. Later that afternoon she approached me and asked, "Where is my headband?"
I said, "I didn't think you wanted it. I sold it for a MILLION DOLLARS."

Her face crumbled and tears began to flow. She was hurt that I would sell her headband for any price. As I was relating this event the next day, I added, "Next time I see her, I will tell her I was joking." Of course in reality, I told her the truth immediately and tried my best to repair my hurtful act. After all I was 'just-picking-on-her.' It was not true and I only said it because I thought it would be funny to see her expression. It was not. This event taught me to take a close look at the potential consequences of a "joke." Experience is the best teacher, as it should be, since it is the most expensive.

The Healthcare bill has ~184,000 words, the Constitution ~4,500, and the Ten Commandments ~313. Sometimes, we do not choose words wisely, and content becomes inversely proportional to volume.

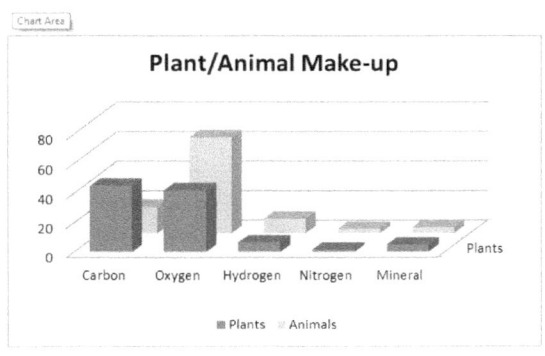

Plants and animals are made of the same stuff.

In order to build something, you must first decide on the materials of construction, i.e., wood, metal, plastic, stone, concrete, etc. The same is true of nature's creations. There are 117 elements from which everything in the universe is constructed. Life on Earth is either plant or animal. The amazing thing to me is that even though they utilize opposite processes (plants: reduction, animals: oxidation), they are both made primarily of the same four elements.

	Carbon	**Oxygen**	**Hydrogen**	**Nitrogen**	**Mineral**
Plants:	45%	42%	6.5%	1.5%	5%
Animals:	18%	65%	10.0%	3.0%	4%

Therefore, animals can eat plants not only for energy but also for elements required to build bodies. For the same reason, some plants eat animals. Is this coincidence or design? It is strange that both life forms on Earth are made primarily of the same stuff.

What are you made of? What defines your Character, Values, Habits, Judgment, Resolve, and Integrity?

All things mineral, plant, or animal are made of some combination of the 117 elements and all are beautiful. *Molly Bawn (1878)* contains Margaret Wolfe Hungerford's most famous idiom: "Beauty is in the eye of the beholder." Beauty could stand as an ordinary miracle because it adds so much to human life. Everything done to excellence is beautiful. Art that is kept for years, decades, and centuries is beautiful. Music that is enjoyed by one or many is beautiful. Nature and the wonders of Earth are beautiful. Lower animals may also appreciate beauty, especially birds. Not only do the males display colorful plumage but the Australian Bowerbird also builds and decorates an impressive bachelor pad to attract willing females. These bowers have a hut-like roof and an entrance that is adorned with carefully arranged pebbles and bright collected objects. This is worthy of a Wikipedia visit.

Horse Feathers

A horse goes into a field, eats grass, and grows hair. He is followed by a sheep that eats the grass and grows wool. In turn, the geese go into the same field eat from the same grass, and grow feathers. We each graze upon the field of life, and each has been given capabilities to produce useful products. Look at yourself: your skills, your talents, and your abilities. Then, produce not what your fellow beings bring forth, but what you have been designed to do.

Hair, wool, and feathers are organic substances and are made of Carbon, Nitrogen, and Oxygen. They belong to a family of amino acid compounds called keratins. Hair and wool are α-keratins which are fibrous proteins. These proteins are also used in the construction of hooves, horns, nails, and claws. Feathers are made of the harder β-keratins which are also used in scales, shells, and claws of reptiles. Porcupine quills and bird beaks are also included in this group. This is an organic chemistry nightmare. Starting with grass, synthesize all of these products and put them in a predetermined location. Do all of the structural requirements, and add special color prescriptions for feathers. I had a Physics professor that wore a bright red tie to class and said he had read that male birds have bright plumage to attract females. He further confided he had no response from women; however, he did find it necessary to fight birds off.

Feathers not only require pigments but also have a variety of structural requirements for the vane feathers of flight. Since feathers are also used for insulation, the down feathers are produced to provide a short soft covering with great density differences between small and large birds. Can this happen without design influence?

As a young person choosing a profession, we often stumble into whatever is available. Then as we work and have income we began to ponder our options. Life would be more productive if we had envisioned early on the profession we would like to pursue based upon our interest, skills, and talents, then prepared ourselves accordingly.

Just as the animals mentioned here each eat from the same field and grow different coatings according to their design, we also were designed with special abilities and should not suppress our given propensities. Feed and develop them into productive and contributing assets.

Vegetables

Can you go into the forest and find a potato? What about a mess of peas, a head of lettuce, or a few ears of corn? Truth is, someone did at some point in time, because we made none of it. Man has cultivated and selectively planted so that the varieties have developed and become abundant; however, the original ancestors of all fruit, vegetables, and berries were given. Most edible fruit have seeds that receive a distribution assist by providing a food value for various animals. This provides the animal world with a great variety of food. There are about 2000 plant species that are cultivated for food. We depend on vegetation for oxygen and food. This is a symbiotic design essential.

There are many advantages to having a diet that is largely vegetable-based. Carbohydrates have half the calories of fats and travel through you digestive system four times faster (12 hrs. for carbohydrates, 48 hrs. for fats). By having half the calories of meat sources, vegetables provide bulk that give you a fuller sensation and encourage calorie intake reduction. Bulk fiber also assists in helping your digestive system from becoming clogged.

Once I was at dinner with a customer. To his surprise, when the ordered entree arrived, it was far more than he could eat. He said, "I cannot eat all of this!" I told him, "Simply eat what you want and save room for dessert." He said, "My grandmother would not allow me to leave anything on my plate and out of respect for her I cannot waste food." He had a principle and stuck to it. This is consistent with the household aphorism of the depression: "Use it up, Wear it out, Make it do, or Do without!" However, a truth learned on the farm is that on planet Earth no food is wasted. If discarded by man, it becomes food for the pets or farm animals. If not eaten by these, this food becomes nourishment for insects. The next step down the food chain is microorganisms and therefore nothing is "wasted." Overeating is unnecessary. There is no need to try to eat all that is available. A system is in place to take care of excess.

Vegetables are a gift. I challenge you to think of the miracle of vegetables the next time you eat a meal.

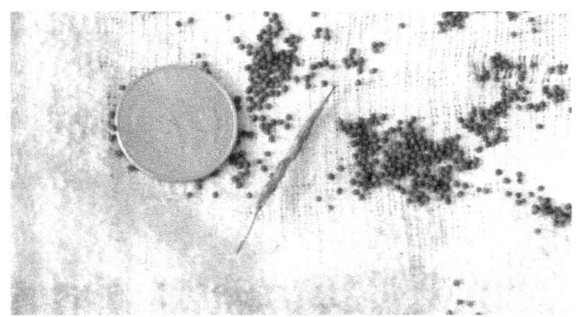
Seeds of Mustard Plant (Brassica juncea)

Seeds

Many plants propagate by seed, like corn, while others, like St Augustine grass, spread by root growth. What a nice design characteristic to have a portable seed to use to cultivate crops! If you consider the information contained in a seed and the potential of that seed, it is astounding. Not only does it contain life to reconstruct its own kind but has with it a library of information about germination, growth, flowering, weathering, reproduction, and so forth. Paraphrasing Luke 17:6: "If you have faith as small as a mustard seed... you can move mountains." (The reference is to the seed of the mustard tree and not the mustard pictured here; however, these mustard seed do possess the same information and are also very small.) The comparison to man's potential has enormous implications. This can be viewed from two perspectives. The first is actual size, and since the seed is physically small the interpretation would be that a small amount of faith is powerful. I can identify with that. The second view is more interesting. Look at the size of the faith possessed by a mustard seed. How much faith does a mustard seed have? If the strength of faith can be judged by action, the mustard seed is loaded. It contains life, a gift that we also have. It waits until conditions are suitable for germination; patience is a virtue that we too can develop. Once the seed starts the germination/growth phase, it gives all that it has to succeed regardless of the environment. If the seed germinates in poor soil, it tries as hard to grow and be productive as the seed in fertile soil by a river. This is a good example for us to emulate. The plant endures the weather: cold, hot, wet, and dry; seldom do we have ideal conditions in our endeavors. The plant bears fruit; so should we. The seeds of the plant contain all the information to produce another generation; we should ensure that our offspring have the same. Thank God for the miracle of seeds and may we use them wisely and emulate their fervor in our life.

Smart Seeds

Many nuts need a cold period, before germination will occur. These nuts mature in the Fall and require four to six months at forty degrees or below and then need a moist shell-softening environment before sprouting. This ensures that they do not start their lives as trees in winter. They wait for the warmth of spring following the winter cold period. We wait until spring to plant our gardens. God plants when the nut is mature with built-in instructions to wait for spring. This can be explained as evolution, i.e., those that did not wait for the cold season to pass did not survive to reproduce like kind. Evolution is an effective tool of design, but actually getting the successful option in position for testing is tricky if left to chance.

This process is called stratification and occurs in many nuts such as: pecan, hickories, and walnuts. Small fruit that require a cold period to break from their dormant state include blackberries and grapes. Some fruit trees requiring stratification are apple, cherry, plum, peach, and persimmon.

This example of a built-in trigger mechanism can be applied to our life by putting more timing into our activities. It is frequently useful to pause before speaking harsh words, especially if you are referred to as one with a 'Short-Fuse.' Counting to ten may not work for you, but you can always go for twenty. It is better to be considered a slow thinker than a hot-head.

Blacksmith Forge Lucedale, MS

Makers

I am thankful for "makers." I ate breakfast this morning with a fork I did not make, from a plate that was made by someone that I do not know. The same can be said about the chair I sit in, shoes I wear, corrective lens, hearing aids, microwave, iPhone, cars, tractors, clocks, and the list goes on. Humankind is unique in the extent that we have the capacity to form a social environment where individuals are rewarded for contributions to the good of the whole. A collective network of people that have the desire, ability, and social freedom to make things is an ordinary miracle. I love and appreciate the way of life my generation has enjoyed and pray for this gift to be extended to my grandchildren.

Makers have brought us collectively to a unique place. I was sitting in the living room one morning when I had a thought that describes our place in time. 'The temperature in the room could have been a couple of degrees cooler to be ideal.' As soon as I had that thought, I realized what a luxury it is to be able to have a thought like that to show up on the radar screen. This is far removed from a survival thought. When we have the ability to be that discriminating, we are reaping the benefits of a great design.

People support other people. We are all interdependent. Our belief in the collective abundance and the mindset of plenty enables us to work and share with the knowledge that there will be more for us as we share with others.

Glass

Can you imagine our world without glass? Glass is everywhere from windows, windshields, light bulbs, coffee pots, stove tops, dishes, and corrective lenses for vision. We even have a wood-burning heater door made of glass to enable a view of the burning wood. Glass is made from sand, silicon dioxide, SiO_2. Silicon makes up 27.7% of the earth's crust exceeded only by oxygen. Window panes are manufactured using the flotation process where molten glass is floated on molten metal to achieve a flat smooth finish. If you do not believe glass is a miracle, then make a windowpane.

When I was a teenager, in 1965, there was an old man who lived in an isolated wooded area a couple of miles from where I grew up. His house was a good three miles down the log-trail road one had to travel to get there; however, a bird could take a much shorter route. I would journey for a visit from time to time, and he was always happy to see me, offering meager refreshments and a look at his carbine rifle which was always rewarding. The first time I found him was when his mule had wandered to my house and I followed him home. One thing that intrigued me about this old gentleman was his simple way of life. His house was made of rough-cut pine and stood unpainted. Nothing was store-bought except maybe rough looking hinges and a door knob. He also had no glass windows. He did have two large windows with shutters that he opened for ventilation, and living near a creek, it allowed a cool breeze to refresh his one-room house. As I am writing this section about glass, I think of Mack's house and realize that his plight was not as good as mine. Some glass, electricity, and a few modern conveniences we now consider necessities, would have made his life easier. We take our world for granted.

Glass is important in our physical world because it is transparent and protective. Is it beneficial for our lives to be transparent? Is that also protective? If your actions match your values and you have no conflict between your decisions and your principles, you are both transparent and protected.

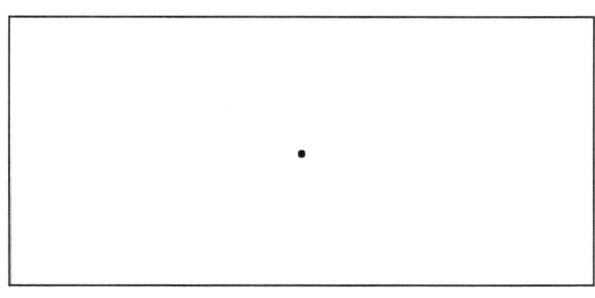

Teachers

Teachers are not made, nor of a chosen few, but they are born. I can say this with some authority because I am one. Not that I am a teacher by vocation, even though I have been hired from time-to-time to teach Statistical Process Control; however, I am a teacher because all people are. Our children are born teaching. From the moment they came into this world they started teaching us the disciplines of parenting. You can attempt to obtain this knowledge elsewhere, but whatever you learn, it is inadequate. If you ever give a presentation, you are a teacher. You have special knowledge of some skill or information that you are trying to impart to others for the purpose of giving them an opportunity to learn. In the animal world, many animals from chimps to ants exhibit teacher activities in passing on needed skills and information to others. They teach by demonstration and/or by providing materials.

John Cotton Dana is quoted as saying, "Who dares to teach must never cease to learn." While this applies to all, it is a trait well demonstrated in our schools and teaching community. Teachers have an awesome requirement to DO IT RIGHT. Young minds need good food to grow, in the forms of nutrition and information. We are teachers and students throughout our lives. We are all connected to all things: plant, animal, material, and spiritual. We communicate at many levels depending on our focus. One cannot walk in the woods and not feel the connection to Mother Earth and the power of Universal Awareness.

Christine Caine is reported to use an illustration of a small black dot projected onto a large white background. When students are asked, "What do you see?" they invariably respond, "A small black dot," failing to notice the other 99.99% of the projection, i.e., the beautiful white background. She encourages us to look at the 'big-picture.' I think this exemplifies a 'Teacher' since they spend their lives opening the eyes of students to the big-picture of their chosen subjects. As students of life, we too are guilty of focusing on the micro without appreciating the big-picture. From the rationale of gravity to the motivation of an electron, the background of design is visible.

Leadership

Human existence depends upon living collectively. We have many systems of networking that link each of us to others. We work together and are dependent upon others not only for convenience but for survival. Our first debt of dependence is to family. How young can a child survive on-their-own? I do not know the critical age for survival; however, in the state of New York a child younger than 12 years of age cannot be left at home alone. We definitely go through at least the first 10% of our lifespan dependent on family support for food, health, shelter, and development. My two offspring were thirty years of age before they were totally self-supporting. This self-supporting age comes earlier if the children do not become medical doctors, but even if they take jobs after high school, it is still a significant amount of time in a take-more-than-give role. In order for human to function collectively, some must assume the task of leader. Leaders are needed for all of the subgroups that we establish whether family, city, state, country, church, schools, business, sports teams, clubs, political party, and on and on. We need leaders to provide a focal point for a subgroup to pursue a common goal, i.e., someone to guide and direct. In the animal kingdom, leaders can change depending on the challenge. For example, elephants follow the oldest female when looking for food and water, but when attacked, group control goes to the dominant male. We are each associated with numerous subgroups. We can be leaders in several and followers in others. This allows humans to specialize and affords freedom to benefit from collective contributions. Highways, economic systems, electricity generation and distribution, telephone, television, automobiles, mass transit, etc. It is interesting that we can have professional sports where the collective pays a handsome price for their entertainment. Such human capacity is beyond extraordinary and exhibits collective society excellence.

Leaders need to step forward. When George Washington served as president, he did several things that set the standard for the office of the president. He chose the name for the office to be president instead of monarch, dictator, king or other lofty titles; however, by choosing the title of president, he elevated it to supreme status. He put in place term limits, and instrumental in solidifying the republic form of government. George was a true leader. We need leaders that seek to provide a system for individuals to be productive and live collectively.

Retirement

What animal other than human can retire? The 2011 Social Security records show 38.5 million retired workers in the United States. With a population of 312 million the same year, that is greater than twelve percent. Think about an animal that reaches an age where they do not have to work to eat. What a social environment to provide such a state! To be able to not worry about your ability to earn your next meal and not worry about a place of shelter or something to wear is a miracle. I know that we do have the homeless and other people that do not have retirement, but the system is in place for those who choose to participate in the work-save-retire process. The beauty of the system is enhanced by retirement not being mandatory. If you are retirement age in the USA (age 65 or older), you are not required to retire. With good health and a willingness to continue to contribute, you can continue working. The wonder of the synchronization of nature around man is astounding, and the social economic structure that it facilitates favors design.

Why should you work when you could become an early government dependent eligible for a handout? Here are a few reasons to participate in work-save-retire process:

1. To grow above a minimum existence
2. Feel good about yourself
3. Be part of the country health
4. The satisfaction that comes with giving more than you take
5. Set an example for your children
6. Develop habits that develop your character
7. Build for your later years

In the animal kingdom, those who do not contribute perish. Look at the drone honey bee. He is not a worker, so in autumn, he is driven from the hive. No need to use valuable resources to feed this nonproductive individual through the winter. We are more tolerant as humans, but a certain amount of pride is worth having. Do not become the subsidized drone of human society.

Light is Something Else

Why is light illuminating? The electromagnetic spectrum ranges in wavelength from thousands of miles as in radio waves to less than the width of an atom in gamma radiation. The thought is that the lower limit is the **Planck length** which is denoted ℓ_P and is a unit of length equal to 1.6.... $\times 10^{-35}$ meters. The upper length for a wavelength is bound by the size of the universe. The portion that we call light has a range of wavelengths from 380 to 760 nm (nm = billionth of a meter). If you lightly touch your fingers together, there is a greater distance between them than the width of the visible light spectrum. If you had to choose a single statistic to support design influence, visible light is a great candidate. The key to the value of this energy is its interaction with matter. This small band of frequencies is the only part of the entire energy spectrum that excites molecular electrons. Excitation of these electrons enables photosynthesis, and gives a wealth of information about our surroundings, i.e., shape, color, and distance. The eye is designed to receive and interpret this information. Light energy is "bright" or "illuminating" only when interpreted by the eye, and then we call it 'visible.'

An additional miracle of light is the sun's energy that reaches the earth is predominately visible light, with a small amount of ultraviolet and some near infrared. With such a range of electromagnetic radiation wavelengths, how difficult is it to recognize design when our sun's output is almost exclusively the visible light? The probability that the sun's output would be this tiny magical band width of only 3.8×10^{-7} meters is a ratio of almost nothing over infinity, $\sim 0/\infty$. The possibility of so much coincidental interaction being associated with this pinch of the electromagnetic spectrum is nonexistent.

Have you ever considered that color is not a physical property? It is perceptual. For example, if you receive light of frequency 660 nm, you see red. The physical part is the ability to receive and identify the wavelength. The perceptual part is your brain generating a color image red. Now here is a thought for you. When my brain generates the 'color' red from 660 radiation and you are looking at the same light, we both agree that it is red. However, it is not a given that we both have the same perceptual image. Your mind's color for red may be my mind's color for blue.

Fire

Fire is defined as rapid oxidation giving off both heat and light. A candle burns at about 460 °F. We use the candle primarily for light and other combustibles for heat. Have you ever thought of how the properties of fire fit our balance of life? The combustion temperature of wood is such that it allows us to heat our homes. Wood starts primary combustion at about 540 °F and proceeds to the secondary stage above 1100 °F. The first two stages are the combustion of gasses before getting to stage three where the slower charcoal combustion takes place. The characteristics of fire are important to maintaining the oxygen balance in the atmosphere. The atmosphere will not support combustion when the oxygen concentration is below 15%. What if it did and wood were explosive or combustion occurred at a lower temperature? The oxygen concentration in the atmosphere has reached an equilibrium concentration of around 21%. The lower limit for oxygen for human survival is 17.5%. Of course we evolve to fit the environment from generation to generation as part of the design; however, it is convenient that we do not have lungs

20 times the size they are now in order to extract oxygen from a lowly oxygenated atmosphere. The ratio of O_2 to N_2 is just right for our form and functionality. If fire was not constrained by the laws limiting combustion, it would be drastically different. Is this design or luck?

The Sun

The sun, made mostly of hydrogen, is the source of energy that drives planet Earth. Ninety-three million miles from Earth and 864 thousand miles in diameter (2.7 mllion miles in circumference) with a surface temperature of approximately 5800 °K (9980 °F), this nuclear furnace has the correct mass to hold the earth at the correct orbital distance so the radiation we receive is perfect for life. Now isn't that a miracle! I have trouble building a fire that will last through the night. God builds one to last billions of years with no life threatening change in view. The mass of the sun is 2×10^{30} kg and it converts 6.2×10^{11} kg (683 million tons) of hydrogen into helium per second. The sun also produces most of the emitted energy as visible light. The sun is such an achievement that it is difficult to take for granted, but somehow we manage.

What great thing would you attempt if you knew that you could not fail? If you could choose one achievement and be guaranteed that you could not fail, what would it be? Be careful with this one because it could be your destiny. When I was asked this one Sunday by Danny Box, a visiting speaker at our church, I immediately thought of understanding gravity. Since gravity is going to be an AAHA!! Moment for someone, why not me? There was a man that felt a strong desire to go into the hills and push against a large rock. After days, weeks, months, and years of pushing against the rock, he became discouraged that the rock had not moved. All of his doubt and negative feelings came into play, and he considered quitting until he realized his objective had simply been to push against the rock, not to move it. By doing this, he had grown very strong and fit. There is nothing that is unachievable. According to Napoleon Hill in *Think and Grow Rich* (1937) your mind will not allow you set a goal that is impossible to achieve. Pick an achievement and go for it. Let your 'Do List' include: believe in it, be diligent, and be persistent; while the 'Do Not List' includes: do not become discouraged, distracted, or dissuaded. You may just surprise the world.

Our sun is a great example of an impossible match of mass, distance, output quantity, and output quality to meet Earth's needs for supporting life in general and human life specifically. This did not just happen by random chance; it is the product of design. Think outside the box to match your activities to your dreams and design your future that defies the odds.

The Moon

The moon is about 1/4 the size of the earth. As the earth orbits the sun it is accompanied by the moon as it orbits the earth. Together, they form a binary pair with an orbital mass center that is about one thousand miles below the surface of the earth. Since Earth's diameter is approximately eight thousand miles, the binary pair shifts the orbital mass center significantly away from the center of the earth.. Without the stabilizing effect of the moon, the earth would wobble and the seasons would be extreme and unpredictable, more like that of Mars. The rotation of the earth is also slowed by the attraction of the moon. A phenomenon known as tidal drag slows the rotation so that the length of a day is 24 hours instead of 10 hours. We are familiar with the effect of the moon's tidal activity, but less obvious is the effect the moon's pull has on land. The earth's crust bulges on the side facing the moon which is believed to be responsible for plate tectonics. This action also generates heat that makes a contribution to the earths mean surface temperature of 57 degrees F.

The theory is the moon was formed as a result of a Mars-size body colliding with early Earth which blasted material into orbit. The material accreted to form the moon. This event is also given credit for tilting the earth axis off perpendicular by 23 degrees. The moon is critical for our planet to sustain life as we know it. How could this not be design?

People too form binary pairs to stabilize life. Appreciate your binary pair partner.

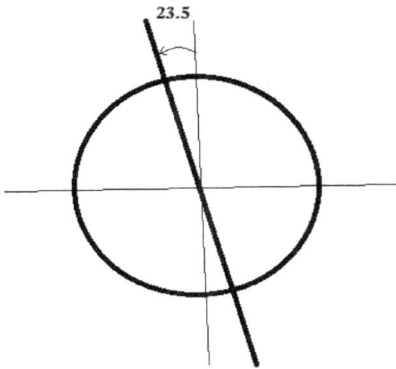

A Tilted Earth

Instead of Earth's rotational axis being perpendicular to its orbital path around the sun, it is tilted twenty-three degrees. As Earth orbits the sun, the amount of energy received by the two hemispheres varies depending on the location along the orbital path. Half of the year, the northern hemisphere is receiving more energy, while the southern hemisphere receives more during the other half of the cycle. This results in seasons, which in turn, make possible life as we know it. If there were no seasons, the equatorial region would receive most of the sun's energy, and the northern and southern hemispheres would get progressively colder with distance from the equator. This would concentrate human habitat to a narrow band near the equator. Since our equatorial region experiences the least seasonal variation, we can get a glimpse of life on such a planet. There are two equatorial region examples: rainforest, like the Congo where the soil is leached beyond supporting cultivation, or arid, like the Arabian Peninsula. Without cultivation, man would be hunter-gatherers in small scattered tribes, if they survive at all. We would have neither wheat, nor corn, nor potatoes, nor barley which grows better in climates that have a cold winter. In addition, man would have to cope with an abundance of virus-carrying insects. Our seasons go largely unappreciated as they sustain our existence. The degrees of axis tilt of the planets in our solar system are: Mercury 0.01, Venus 2.6, Earth 23.5, Mars 25.2, Jupiter 3.1, Saturn 26.7, Uranus 82.2, Neptune 28.3, and Pluto 60.4. Are we lucky or is this an ordinary miracle?

All planets are tilted to some degree. So are all humans. Look at the 'tilt' of your life and keep it in check while being tolerant of others whose 'tilt' may not be as ideal as yours.

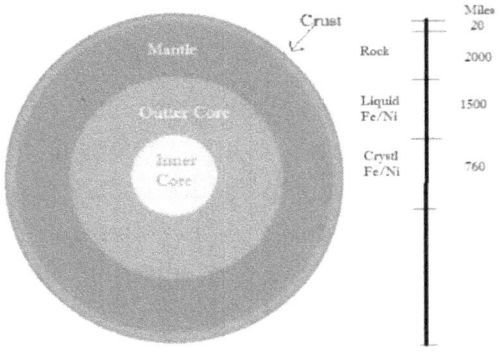

Magnetosphere

The earth comprises four distinct layers: Crust, Mantle, Outer Core, and Inner Core. The Outer Core is below the 2000-mile-thick Mantle and is mostly molten Iron. At the center of the earth is a solid metallic crystal Inner Core that is 760 miles in diameter, i.e., 70% the diameter of the moon. The Inner Core spins independently from the spin of the earth. The earth's magnetic sphere, the magnetosphere, is produced by the interaction of the Inner and Outer Cores. The significance of this magnetic field is that it shields the atmosphere from being stripped away by the charged particles in the solar wind. The solar wind is a steady stream of 250 miles per second of charged particles. The magnetic field of the earth extends ~3 million miles on the side of the earth that is away from the sun and is compressed to ~40 thousand miles on the side facing the sun. Why does Earth have a magnetosphere when Venus and Mars, our solar neighbors, do not? As a result they have no appreciable atmosphere. Go figure.

Ozone Layer

Oxygen is most common in the atmosphere as O_2. That is a molecule of two oxygen atoms. Oxygen also exists as Ozone, O_3, but in a much smaller concentration than O_2. Only 0.6 parts per million O_3 is found near the surface. However, there is a heavier concentration, ~10 ppm, in the lower stratosphere 12 to 19 miles up. This layer of ozone absorbs nearly 98% of the solar ultraviolet medium frequency radiation (200 – 315nm) which would be detrimental to exposed life on Earth's surface. A great number of odd occurrences are needed to support life. Are these fortuitous occurrences or design? I remember a *Star Trek* episode where the Enterprise had just safely navigated through a heavily clustered meteor field when Spock said, "Random chance has operated in our favor." McCoy responded, "Random chance had nothing to do with it. We were lucky." Spock replied, "I believe that is what I just said." The evidence goes beyond luck. Random chance would not cover all bases.

O_3 is present in a small concentration with respect to O_2. The ratio is 1/21,000; however, this small component is critical to life on Earth. Small acts of kindness, sharing, and caring from your life can be significant and is important to the health of the whole.

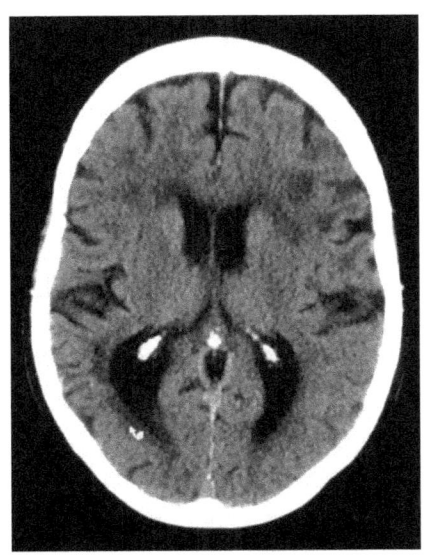

The Human Brain (hardware)

How many gigabytes is this biological computer that we call the human brain? I saw a guess that went similar to this: "The human brain contains ~ 100 billion neurons and each neuron has ~ 1000 connections. This equals 100 terabytes if each connection is equivalent to one byte of memory. So, in gigabytes that is 100,000." Actually, this is a huge understatement because the brain does not store in static charges. It works differently using currents and chemical storage. The brain is taking data from all five senses (sight, sound, taste, smell, and touch), analyzing processing and storing for a lifetime.

On October 20, 2013, Ann and I attended Homecoming at Shipman United Methodist Church (pictured on the cover) and had a reunion with a high school friend and the Best-Man at our 1966 wedding. The interesting part of this encounter is the memories from that era came vividly alive. We discussed our early exploits in a 1951 Ford convertible in some detail. No written record, just information tucked away in the brain for recall on such an occasion as this. The memory trip included a one-night venture to a swimming-hole located a half mile off the main road up a logging trail. The main road was not much better, a county dirt road that was graded once a month at best. We were in the convertible that had poor-to-none brakes and a huge V-8 engine. As we drove the logging trail, we encountered a twenty-foot-wide mud-hole where I over-accelerated to avoid getting stuck. The operating philosophy was, 'As long as you are moving, you are not stuck.' I still find this philosophy applicable to many endeavors. However, on this occasion, I lost control after successfully avoiding getting stuck, and we became wedged between two trees. When the car was free, we found the swimming-hole in the very cold Rocky Creek and used the car headlights to see how to climb the tree used as a diving platform to enter the dark water below. This event, along with many others, was safely stored in two brains. The information is available in visual, tactile, auditory, olfactory, and gustatory memories. To have a record of a lifetime is an awesome amount of data.

The Human Brain (software)

As impressive as the 100 terabytes of capacity the human brain has as hardware is, the software side is even more impressive. Our biological computer comes genetically loaded from the experiences of our ancestors. In other animals, we call this instinct. An example is a bird that is booted from the nest, as it matures to flight-capability, is able to fly without training. This same bird is able to build a nest just like mama when it is an adult. I was raised on a farm and have wondered how a newborn pig knows where to go for his or her first drink of milk. Within a minute of birth, the piglets are making their way to mother's udders and the teats toward the front have the most milk, which is why the first ones out go there. Starting from a single cell that has the information to build an entire body and furnish genetic memory information for survival is miraculous.

In the 1970 movie *Patton,* star George C Scott said, "The Carthaginians defending the city were attacked by three Roman legions. The Carthaginians were proud and brave but they couldn't hold. They were massacred. Arab women stripped them of their tunics and their swords and lances. The soldiers lay naked in the sun. Two thousand years ago. **I was here.**"

I had a personal genetic memory event. When Larry Jr. was five years old, he was 'helping' me replace the oil pan on a small pickup truck engine. In order for him to have something to do, I started the pan bolts and his task was to screw them in and tighten with a ratchet. As he proceeded with this task he asked, "How tight should I get these? The last time that I did this I rung one off." We both looked at each other in amazement of the comment. He had never done this before. However, before he was conceived, I did ring one off working on the car Ann and I desperately needed to use to get to work the next day. The removal of that broken bolt was a serious task that I remember well. The car was ready to go the following morning.

My daughter and I communicate very well. I can start a sentence, and she can finish it. I will be trying to think of someone's name and will say something like, "Who is the guy that ...." and she will interrupt and say the person's name before I get enough information out to narrow the candidates from the population at large. The human brain software is a miraculous piece of work.

We also have a very capable self-programming function. We do not casually rewrite code. This caution gives birth to superstition. Until we are sure the rubber band we wore on our tennis shoe was not responsible for our outstanding athletic performance, we continue to wear it, not wanting to change anything that is working. Reading this document and contemplating the validity of concepts set us apart from all other creatures.

Evolution

Evolution has always been falsely considered the enemy of creation. Actually, evolution is one of the miracles of creation. We not only evolve physically over generations but mentally throughout a lifetime. We can see change. The average American in 1850 was 5'7" compared to 1980's 5'10". All living creatures are changing and adapting to the changing environment. A survivor from George Mallory's crew returning from their 1922 third failed attempt to reach the summit of Mt Everest, the world's highest mountain, addressed a welcoming crowd by speaking to the mountain: "You defeated us on our first attempt, again on our second, and yet again on our third, but you cannot win. You are getting no bigger, and we are." Growth and change are an important part of our design. Survival over the long term depends on evolving to better cope with prevailing conditions.

Man's capacity to learn is also growing rapidly. Read the work of James Flynn, a professor *emeritus* at the University of Otago in New Zealand. He points out IQs rose seven points per decade from 1952-1982, and the rate is probably not linear over time. Since evolution is going to happen, we should endeavor to nudge those within our sphere of influence in a positive direction, while defining positive as that which is conducive to the survival of humankind.

Genetic Engineering

Great White Sharks do not cruise rivers and lakes. This is a good thing for humanity. They are restricted to saltwater because osmosis effects keep freshwater-fish in freshwater and saltwater-fish in saltwater. Slightly less than half of fish species are freshwater (41%) and the remaining are saltwater. The two groups are physiologically different. Saltwater fish have tissue that is less dense than their surrounding environment and the salt water is constantly extracting water that diffuses through their skin and as a result they drink water constantly. Freshwater species require low salinity water, usually less than 0.05 percent, are denser than surrounding water and their bodies absorb water. They do not drink at all, but are continuously urinating to remove excess water. Freshwater fish differ physiologically from salt water fish in several respects. Their gills must be able to diffuse dissolved gasses while keeping the salts in the body fluids inside. Their scales reduce water diffusion through the skin. The kidneys of freshwater fish are well developed to reclaim salts from body fluids before excretion.

There are a few varieties that can do both. Salmon have an anadromous life strategy and spend most of their adult life at sea, returning to freshwater for reproduction. Some species, including the bull shark, can survive both environments; however, evidenced by their large abundance at sea verses their rarity in rivers, they are better suited for a high salinity environment. The different physiology for saltwater vs freshwater fish is a product of genetic engineering, which is a wonderful tool, where surviving species have stood the test-of-time.

In 1978 Genentech, the first genetic engineering company announced the production of genetically engineered human insulin. The insulin produced by bacteria, branded humulin, was approved for release by the FDA in 1982. This was, and is, a great event that gives a much needed product to the diabetic community. This presents a dilemma. The test-of-time is skipped with human intervention. In the natural process of evolution thousands of years are spent in the genetic modification process, weeding out the ill-fated changes. With humans short cutting the process, the test-of-time comes in the distant future and the journey has begun with no certainty of the consequences. While the bacteria production of human insulin is welcomed by most, other genetic modifications may not be as well received.

A joke told by Ronald Reagan in 1985 demonstrates the issue, and I use it because only the test-of-time will reveal the outcome of actual genetic alterations.

> "But this is the story of a realtor who was out driving on a back road on his way to look at some property and suddenly noticed down beside him was a chicken keeping pace with him, and he was doing 60 miles an hour. And suddenly the chicken spurted out ahead of him. And it looked to him as if

the chicken had three legs. And then it turned and went down a side road and into a barnyard. And the driver turned down that lane, drove into the barnyard. There was a farmer there and he asked him, 'Did you see a chicken go by here?' And the farmer says 'Yep.' 'Did it have three legs?' 'Yep. I raise them that way. I breed them.'

Then the realtor asks, 'You do? How come?' 'Well, I just love the drumstick and Maw always liked the drumstick and now Junior's come along and he likes it and we just got tired of fighting over it. So I've been breeding three-legged chickens.' " The realtor then asks, " 'Well how do they taste?' " And the farmer replies, " 'I don't know. I haven't been able to catch one yet.' " *New York Times, April 20, 1985*

Caution!!! Our well thought-out genetic design changes could have undesirable long-term side effects.

The Speed of an Electron

All chemistry students are familiar with the description of the hydrogen atom where there is a proton at the center with a single electron in orbit. A first time photograph was taken of a hydrogen atom using the newly developed "quantum microscope" by Anela Stodolna of the FOM Institute for Atomic and Molecular Physics in the Netherlands. The picture showed the electron cloud that is consistent with the Schrodinger equation (a partial differential equation describing the quantum state changes with time). How does one get a cloud from a single electron in orbit? To form a cloud instead of a ring requires critical directional changes. We are getting smarter all the time, but still not smart enough to explain this design. What motivates the electron? It travels at the speed of light. What keeps it going? Like gravity this will require an AAHA!! Moment.

Motivation of an electron is more elusive than motivation of humans. We are motivated by a variety of sources including other people. Many years ago, I had a brief encounter with a young man who told me something that has ever since been etched into my mind. The lad was a Filter Stripper in a viscose rayon plant. The filters were huge plate-and-frame design and the work replacing the filter cloths was a nasty job that required heavy lifting and working in a caustic environment. Few workers were able and/or willing to perform this task. I am sorry I do not know the young man's name; we will call him John. As we passed in the corridor outside the filter area, aka 'The Cave,' I asked John, "Are you having a good day?" His reply has been a truth that I have repeated often. He simply said, *"Life is too short to have a bad day."*

Enzymes

Chemical reactions that routinely take place in life are impossible in the lab. This could not happen without enzymes. Enzymes are catalysts that are highly selective and greatly accelerate both the rate and specificity of metabolic reactions. From the digestion of food to the synthesis of DNA, enzymes are required. In order for enzymes to work, the temperature and pH must be just right. Like all catalysts, enzymes work by lowering the activation energy for a reaction, thus dramatically increasing the rate of the reaction. As a result, products are formed faster and reactions reach their equilibrium state more rapidly. Most enzyme reaction rates are millions of times faster than those of comparable un-catalyzed reactions, and some are billions of times faster. As with all catalysts, enzymes are not consumed by the reactions they catalyze; neither do they alter the equilibrium of these reactions. However, enzymes do differ from most other catalysts in that they are highly specific for their substrates. Enzymes are known to catalyze about 4,000 biochemical reactions. In the conversion of N_2 into NH_3, a form usable by plants and animals, all bacteria use the same enzyme, nitrogenase. Is it just me, or is speeding up a reaction a billion times miraculous? In industry we looked for a project to yield at least a 10% improvement to be significant. Another interesting fact is enzymes are required to make enzymes. This makes the question of, 'Which came first, the chicken or the egg?' seem trivial.

Hemoglobin

Hemoglobin (Hgb) is an iron-containing protein that carries oxygen in the blood of all vertebrates, with the exception of the ice fish of Antarctica whose blood is near freezing and O_2 dissolves better at lower temperatures. Their blood is clear instead of the customary red that is imparted by hemoglobin. In all other vertebrates, the presence of Hgb increases the capacity of the blood to carry oxygen by 70 times compared to oxygen solubility in the blood. One Hgb molecule binds four O_2 molecules making it very efficient. The trick is that once reaching the cell, Hgb must release the O_2 where it can be used by the organism and at the same time bind the cell's waste CO_2 for transporting out of the body. It accomplishes this by having more affinity for CO_2 at the high pH in the cell and more affinity for O_2 at the lower pH in the lungs. This could be considered a pretty slick trick of coincidence or another ordinary miracle of design.

Glucose/Ketones

The body runs on glucose and some ketones. When you eat carbohydrates or protein the body converts them into glucose to be burned in the cells and stores the excess in body fat. This is equivalent to having an automobile that runs on any fuel that contains the correct atoms by having a processing plant built in to convert what is currently being put in the tank into gasoline. When you eat fats, you get fats and cholesterol.

In organic chemistry, I remember problems like, "Starting with acetylene synthesize heptane." We never had a problem like, "Starting with a turnip root synthesize glucose and some fatty acids." The reason of course is professors had to stick to problems they could actually solve. The body routinely converts a wide variety of food intake into glucose.

Now an interesting thing is when the supply of carbs and protein is low the body uses the stored fat for energy. This process breaks down body fat into ketones which are burned preferentially by the brain and heart. These organs are high on the priority list to sustain life. The brain uses a great deal of energy. While the brain is only 2% of our body weight it uses 20% of the body's oxygen and 10% of the stored glucose. Ketone burning, ketosis, in the brain is more efficient and essential for a healthy brain. Ketosis is induced when glucose concentrations are low.

Babies on mother's milk are frequently in ketosis which promotes healthy brain growth. Low carbohydrate diets are prescribed for epileptic patients and other brain disorder patients to induce ketosis for the beneficial effects. At night our bodies cycle to mild ketosis if carbohydrate intake is normal, and this is considered a necessity for brain health. What a miracle to convert our wide range of food intake to glucose and store the excess in fat cells to be used later. What is the difference in excess and surplus? The energy available from too much food in your stomach is 'excess' and the energy in fat is 'surplus.'

The Eye

Darwin said explaining the evolution of the eye would be difficult using his new theory of natural selection. Evolutionists say difficult, yes, but not impossible. Zoologist Dan-Erik Nilsson explains how this complex organ could have evolved starting with a light sensitive patch on the skin and given 365,000 years to fine tune. The skeptics say that there are too many independent parts of the eye that have no supporting thread to evolve. How could the lens, cornea, pupil, iris, and retina evolve when each are useless without the others? Well easy enough if you consider evolution with a design influence. Science is the study of design details and evolution is one of the tools. Not only is the eye itself miraculous, the usage and availability of visible light with its unique interaction with matter points to design. While any one ordinary miracle may leave you on the fence teetering between random chance occurrence and design influence, collectively they speak volumes.

The debate of the origin of the quote, "The eyes are the window to the soul" continues; however, there are scientists who contend that the statement is true. Everyone has a different structure of lines, dots and colors in their iris. Researchers at Orebro University in Sweden found that people with densely packed crypts are more warmhearted, tender, trusting, and likely to sympathize with others, while those with more contraction furrows were more neurotic, impulsive and likely to give way to cravings. They think that eye structure and personality could be linked because the genes responsible for the development of the iris also play a role in shaping part of the frontal lobe of the brain, which influences personality. The findings could one day be used in psychoanalysis and by companies screening candidates for jobs.

The human eye is a great example of design influence.

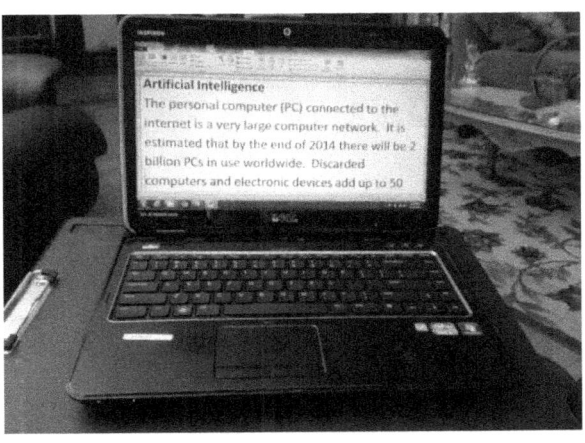

Artificial Intelligence (AI)

The personal computer (PC) connected to the internet is a very large computer network. It is estimated that by the end of 2014 there will be 2 billion PCs in use worldwide. Discarded computers and electronic devices add up to 50 million tons annually of electronic waste. With this volume of computer hardware available to become connected through the internet, the resulting collective brain would be enormous, if it was intelligent.

One Sunday our pastor made a comment about Johnny Appleseed and related his 1800s travels across the country planting apple seeds. He also mentioned that he did not recall his given name. Three people had the answer in less than a minute with an iPhone internet search: Jonathan Chapman (1774-1845). We also knew his birth place, where he died, his parents, he was also a missionary for The New Church, and more. This is impressive; however, it is not intelligence. It is difficult to think of a question for which an answer cannot be retrieved if anyone has recorded such on an internet-connected source. With all of this connectivity and sophisticated programming, why isn't the internet intelligent?

Reading the history of personal computers online in Wikipedia **Personal Computer** will give you a good idea of the exponential growth of the power of computing since the mid-1960s. The growth has been phenomenal and the next fifty years are beyond our wildest dreams. Will we have brain implants that give thought access to internet information?

What would it require to make the internet intelligent? Many computer systems fit the common definition of Artificial Intelligence that was coined by Russell & Norvig (2003), i.e., "A system that perceives its environment and takes actions that maximize its chances of success." This is far short of human intelligence. The human brain has the ability to rewrite code, i.e., modify software. That is the essence of intelligence. To be able to experiment and based upon outcome change the software. When this happens in computers, they will be intelligent. Self-aware is another topic.

Love

What branch of science is love? Psychology and Social Science are good candidates. You can choose; however, it is an important part of human survival. Wikipedia states, "In terms of interpersonal attraction, four forms of love have traditionally been distinguished, based on ancient Greek precedent the love of kinship or familiarity (in Greek, *storge*), the love of friendship (*philia*), the love of sexual and/or romantic desire (*eros*), and self-emptying or divine love (*agape*)."

A definition of love that captures the magnitude of the word is, **'Love is the willingness of one to share their life with another.'** The sharing of life can be in any form, i.e., food for the hungry, money for the needy, clothes for the disadvantaged, time for the distressed, shelter for the exposed, affection and touch for your lover, and many more forms.

The love experienced by a couple that blend mind, body, and spirit making the two into one being is overwhelming. A sustained feeling of completeness is shared. The love expressed in giving always rewards the giver as much as the receiver.

The drawing on the previous page is interesting in connection with the human emotion of love because it physically demonstrates a psychological need for completeness. The triangle appears to be on top of three black circles. If the triangle is on top of the circles, then it must be closer to us than the background and therefore looks whiter. The triangle, however, is not part of the drawing. It is a manifestation of the mind to complete the picture. You can visualize a lines that goes between the circles that complete the drawing. The drawing is simply three black circles with slices removed.

An interesting example of sharing happened in the early 1980s when my daughter, Leigh Ann, was 4 years old and John Schneider was at the Mobile Municipal Auditorium for a car show featuring the *Dukes of Hazard* famous car, i.e., *General Lee*. Leigh Ann was a big fan and wanted to go see 'Bo Duke.' We took her to the show, paid the entry fee, looked at the cars, and then proceeded to try and find Bo. As it turned out, he was on a stage sitting behind a table autographing pictures, and we received a numbered ticket that was near 1100 on the list. We were at the end of the line with approximately a thousand people ahead of us. After standing in line for about ten minutes, the reality of the situation became obvious, i.e., we were not going to make our way to the stage before Bo left. I told the family to follow me. We went to the head of the line. As we left the end of the line starting toward the stage, a man told me that they would not let me on stage until my ticket number. Once we were at the stage, I offered the next person to go up $10 for his ticket. He declined as did the next five people. The next person was an elderly lady that had two tickets. She had been in line for hours. She looked at Leigh Ann and said, "I don't know why I am here. I don't care anything about going up there. You can have these tickets, and I do not want your money." I know why she was there. The love expressed by this lady was apparent as she shared her life with my daughter.

Man could not survive without sharing lives with the human collective?

A Love Song

Poets have tried to say it, "I love you."
Writers through the ages have told it too,
Singers have sung it, that their love is true,
But I want to live my love to you.

Love can't be bought, it can't be sold,
It floats on a market much higher than gold.
The more love you give, the more you are due.
I want to live my love to you.

I love you is such a simple little phrase,
It gets said through life in so many ways,
But they all have in common this thought if they're true,
I want to live my love to you.

Written by Larry Henderson, Sr. and sung to Elizabeth Ann (Baygents) Henderson. It must have been really bad. The first time I sang it, she cried.

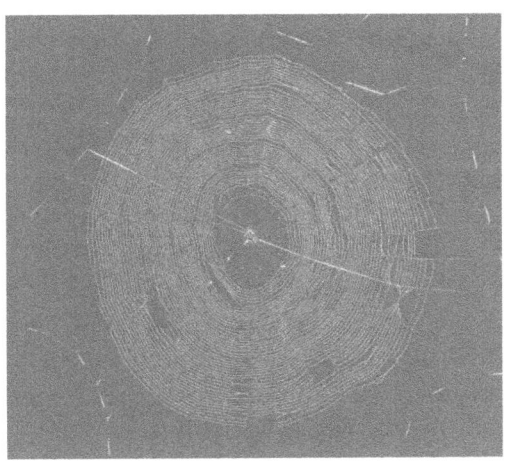

Difficult to photograph the invisible

Spider Web

The spider web does not meet the criteria for an *Ordinary Miracle* in the sense it is not apparently essential for life on Earth; however, it is such an obvious product of design, it merits attention. An article in the August 2014 issue of Readers Digest drew my attention to Biomimicry, a consulting firm that works with major corporations as they look to Earth for nature's answer to engineering problems. Owner Janine Benyus explains, "If something can't be found in nature, there's probably a good reason for its absence."

The spider web held the answer to an increasing problem of birds being killed by flying into windows. Modern architecture utilizes large amounts of glass which creates a huge problem for birds. We had a large glass-enclosed staircase on the end of our house for over 20 years, and each year several birds would collide with a window and die. Seeking a solution to this extensive issue, Biomimicry observed that spiders build webs for bugs, and it would be disastrous to web construction and the livelihood of spiders if birds errantly flew through the webs. When I first read this, my mind jumped to the conclusion that they were going to recommend screens for the windows. I was wrong. The answer is much more subtle. Spiders weave an ultraviolent-reflecting component into their webs. Bugs and humans can't see it, but birds can. I am impressed, and I do know that the spider did not think of this. Something so odd and effective as an ultraviolet-reflecting component being added to the web, accompanied by the bird's exclusive visual spectrum range to include the ultraviolet region, has to be part of design. Now ultraviolent-reflecting components are added to bird-safe windows. I wish Biomimicry the greatest success in incorporating nature's proven design parameters into human engineering.

Life

Evolutionists try to find a spark of life outside of Earth to signify life is spontaneous in the universe and happens without design influence. If life is found in space that did not originate on Earth, it will only prove that the universe is God's laboratory. When you consider what life is and what it takes to sustain it, the probability of random chance development is remote. Life is defined in Wikipedia as *"A characteristic that distinguishes objects that have signaling and self-sustaining processes from those that do not, either because such functions have ceased (death), or else because they lack such functions and are classified as inanimate. Biology is the science concerned with the study of life."* One of the characteristics of the self-sustaining process is called homeostasis, i.e., the property of a system that regulates its internal environment and tends to maintain a stable state. I will not attempt to look at all the complexities of life, but just enough to show the probability of design. Homeostasis is a good place to start. Homeostatic systems in the human body include: Circulatory, Lymphatic, Nervous, Endocrine, Respiratory, Digestive, and Urinary. These are all independently critical to maintain the stability in the body that is required for life.

Circulatory hemostasis is critical and complex. Delivery of nutrients and removal of waste products are the essence of the circulatory system. Nutrients pass from capillary blood into fluids surrounding the cells as waste products are removed. This system must regulate the diverse functions and component parts of the cardiovascular system in order to supply blood to specific body parts according to need. It also ensures a constant internal environment surrounding each body cell regardless of differing demands for nutrients or production of waste products. WOW! And we can take this equilibrium for granted.

Lymphatic hemostasis is carried out by performing three main functions: 1) fluid balance, 2) fat absorption, and 3) defense. Fluid balance is achieved by the lymphatic system taking up excess tissue fluid and returning it to the bloodstream. Fat absorption is achieved by the lymphatic system absorbing fats from the digestive tract and transporting them to the bloodstream, and defense is achieved by the lymphatic system helping to defend the body from disease by utilizing white blood cells. The lymphatic system provides the lung tissue in the bronchi with a mucus layer and helps the respiratory system by creating mucus in the nose to filter out allergens.

These are two of the five support systems that are continuously providing balance and equilibrium essential for human life. This is too much development of independent control systems to not have design influence.

Integrity/Relativity

Much of relativity is based on the speed of light being constant with respect to the observer which makes time a variable depending on the motion of the source with respect to an observer. Einstein had two connected theories on relativity: special relativity and general relativity. The theory of special relativity (1905) applies to all physical phenomena except gravity. The general theory (1916) provides for the law of gravitation and its relation to other forces of nature. Special relativity ultimately replaced Isaac Newton's 200-year-old theory of mechanics. Mindsets changed slowly. Special relativity was generally accepted by 1920 and general relativity, due partially to the complexity of the associated mathematics and also because most applications were for astronomical scales, was not embraced until 1960, over forty years later. The essence of all things being relative is now obvious and whether the speed of light is constant or variable, which is still a debated topic, all things are relative, even time.

Life passes slower to a young person than one that is aged. One hour represents a larger percentage of one's life as a young person than it does in an elderly individual; therefore relative to their exposure to time, the perception of an hour is longer to a young person. One of the rules of 'Time-Out', as a corrective measure for children, is to make the child sit in isolation for one minute for each year of age. One minute, to a one year old child, is as long as two minutes are to a two year old. Have you ever thought about a housefly's ability to escape as someone would try to catch it? Is it because their reflexes are faster than human, or because the fly's short lifespan gives it a different perception of time so that it sees us moving in slow motion?

Rich and poor are also relative. The very poor do not require much to be wealthy while the very wealthy have little reference to daily needs of food or shelter. How we see life and others is biased by our own environment. How we see ourselves is also relative to our core principles and values. One of the measures of a developed character is integrity. Integrity is always respected and held as an essential for corporative living. Not only is integrity respected in others, but it is the primary source of self-respect.

In the mid-1980s, I went to Germany to view a new piece of equipment, and was accompanied by my family. While I was at a manufacturing site during the day, Ann, Larry Jr., and Leigh Ann shopped in two nearby towns: Blumenthal, and Vegasack. A marked difference between shopping in the USA and these German towns was the trust of the German merchants. Ann had no shopping bag and had purchased several items that she carried in her hand. As she entered the next shop, she asked a clerk, "How will you know what I pick up in here since I have in hand these items?" The clerk laughed and said, "Surely you will know what you pick up in here." It is so good to have reason to expect people to do the right thing.

People are more honest when the honor system is used. I sold a Pick-up truck load of watermelons from a parking lot via the honor system. I had a sign and a bucket. The sign read, "Watermelons $2/each. Place money in the bucket." I was told that when I returned that I would have neither melons nor bucket. This was half correct. The melons were gone; however, I had three dollars more than due.

I did something as a young man that I remember with pride. It made me feel good and has remained as a reminder that integrity rewards. For me to have a building project that required a visit to a local sawmill was not unusual, and this was such a day. I went in, placed my order, paid, and left. As I was returning home with the lumber, the price did not feel correct. At sawmills, the costs are computed in board feet. Everything is sold as if it were 1 x 12 cuts. If you buy a 1 x 4 the length is divided by 3 to give the 1 x 12 equivalence. When I computed the cost of what I had purchased, the amount that I had paid was significantly less. I went back to the very busy sawmill and took the ticket in to Mr. D and told him that there was an error in the bill. He was extremely frustrated, snatched the ticket and said, "That is the trouble with people who do not understand how to figure board feet!" He then began to compute, with pencil and paper of course. When he finished, he said, "There was an error alright, YOU OWE ME $105 MORE!" I told him that is what I thought and had returned to pay. As I was near the door on the way out he said, "Hey, I'm sorry. I did not expect that." Integrity is so essential and still we do not expect it in others, neither do we strive to perfect it in ourselves. The term "Honor among thieves" is derived from selective integrity. Be honest and relatively speaking, it will serve you well.

Humankind thrives on integrity. Collectively society must have trust in order to function, and integrity is the miracle that makes trust possible. In industry, one of the no-tolerance-rules is that of integrity. For example, falsification of records is grounds for immediate dismissal. The miraculous thing about truth is, as humans, we sense it. When you hear a new 'TRUTH,' it resonates with you as if you already knew it. I am sure you have found examples of this while reading this book, just as I did while writing. All things are relative; however, integrity is an exception. It is as close to an absolute as you will find. Embrace it as a principle and let your word be your bond. Matthew 5:37 – "But let your 'yes' be 'yes', and your 'no' be 'no'."

Antarctica

The coldest temperature ever recorded on Earth, *-129.3°F*, was recorded in Antarctica. It is the coldest and windiest place on Earth and home of the South Pole. Wind speeds reach 200 miles per hour. Ninety-eight percent of the land mass is covered by a mile-thick layer of ice. The annual precipitation is about 8 inches along the coastal regions and far less inland. While there is no permanent human population, there are about 4000 seasonal scientist and 15,000 tourists annually. It is home to the Emperor Penguin, but due to the inhospitable environment, they are pretty much alone.

The miraculous contribution of Antarctica to our existence is that of a temperature regulator for Earth. The Southern Ocean is a significant sink for both heat and carbon dioxide, acting as a buffer against climate change. Sea ice forms around the continent each winter and controls the exchange of energy between the sun and the earth. When sea ice forms, salt is rejected from the forming ice and increases the density of the upper ocean. These waters then sink and form the deep ocean currents that carry heat around the globe. Remember those deep ocean currents from the *Energy* section?

Antarctica also regulates planet temperature by reflecting a large amount of the suns radiation. The large sheet of ice covering Antarctica gives 5.4 million square miles of white surface area to deflect solar radiation thus making a significant contribution to Earth's thermal equilibrium.

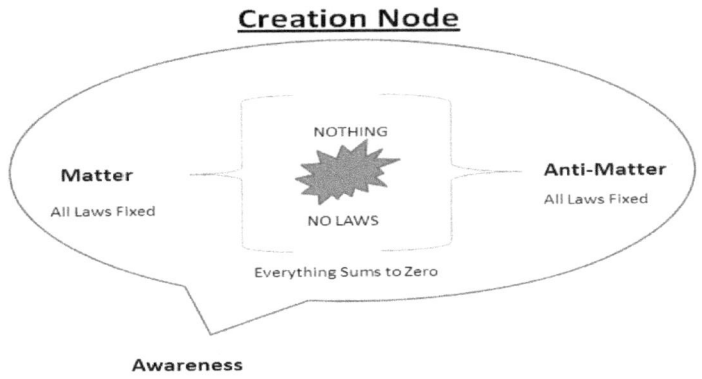

Who is God and where did He come from?

"In the beginning there was God." That is easy to say, but where did God come from? What was His beginning?

Before the beginning there was nothing. Nothing is an ideal place to begin because the possibilities are neither bound by rules nor laws. There are no rules, laws, or time. Time is distance divided by velocity. If there are no particles to measure distance between, then there is no time. When I say 'nothing' that means there was only void without bound. Is empty space nothing? Not anymore, once there are two particles everything becomes something with respect to those particles. The particles would be some distance apart, moving at some velocity with respect to each other, and now with speed and distance, we also have time. Before two particles, there was no Time, Physics, Chemistry, Geology, History, etc. E does not even equal MC^2 yet. All of this happens when we have two particles. What the particles are and how they behave with respect to each other are the laws. With creation come the laws that are identified by observing the output of creation. Different creations would have different particles and different laws. Can we start with nothing and end up with all that we observe? Unless you believe in hocus-pocus the answer is yes. Just look around you. There had to be a beginning. The famous British astronomer Fred Boyle declared that there was no beginning and the universe is 'steady-state.' We only think of a beginning because we are finite and matter has always been here. I am not from that school. To me, there was a beginning and this is how I see it progressing.

A nothing state is so rich in possibilities. No particles, no energy, no rules. Suppose two things popped-up and if put back together equaled zero. You could have anything as long as they cancel each other out. Remember, no rules, so you cannot say there is no energy to cause such an event. There are no rules before the event. Anything can happen if the net sum is zero. Suppose all over the nothing void opposites are blinking up only to immediately cancel each other out. Then two particles blinked up that repelled each other and moved in opposite directions. This repelling and possibly explosive escape velocity could stimulate a sustained creation node producing matter exploding in one direction and anti-matter in the other. If the direction of travel were semi-random, the total creation node would be a sphere with the matter half going one direction and the anti-matter half going the other. The rate of generation would be born with the creation node event as well as all the laws of physics and nature.

Now suppose that along with this creation node there was an accompanying Awareness. Many philosophies include a spiritual side to man. Is this an attempt to explain the unexplainable with a higher power or is it the result of an enlightened mind observing this creation? If the Awareness did accompany creation could it possibly do anything with matter? Mind over matter demonstrations by man have been nonproductive. Always some trickery is involved. Yet all of our individual lives have been touched by a sense of spirituality. While working in manufacturing, I trusted the CEO when directional choices were made because he was in a position to see for the whole of the organization and not just the manufacturing sector. He had more data, more contacts, and in general a big-picture perspective. The same is true with the universal Awareness, except exponentially multiplied: Intelligence on steroids. The creation node could be producing matter at a rate of a universe per second or much slower. One can imagine this half sphere expanding for a time and gravity pulling enough matter together for a big bang and birth of a new universe subset.

Coexisting with our creation node in some distant section of the void could be other nodes that have different matter with different laws. The only common thread is that each nodes output sums to zero. The Awareness at first would be observing then understanding, then influencing, then manipulating, followed by organizing and directing. One big bang followed by another; and within our half of the total sphere, and there could be multiple big bangs taking place substantial distances apart. This would give opportunity for mastering the art of directing the flow of energy and matter, i.e., to choreograph a perfect planet.

What evidence do we have that this Awareness exists? All the ordinary miracles are a good indication. Things that appear impossible without design influence are probably the product of design. Can I move a stone with thoughts? Not yet. I have never seen anyone with telekinesis ability; however, I do believe that we are all connected to the universal, infinite Awareness. The fact that great minds have expressed the postulate that the mind affects reality tells me if you multiply this by infinity you have an Awareness that can. I also believe that being connected to the universal Awareness is not limited to our physical body but is freed at death to observe the universe. I have loved ones that have entered that realm and believe their awareness is still with me.

Conclusion

Life on earth has an impossibly low probability of occurrence without design influence. This is true even if you only consider the essential variables that we can easily identify and semi-understand. To put this in prospective, consider this analogy.

What is the probability of a house light bulb working when the switch is placed in the 'ON' position? To be a little more specific we will use your house and the light by your front door. Further, we will say that it worked as expected last time it was tested. The expectation is when the switch is placed in the on position the light will come on. The probability is very high. A probability of 1 is 100% so we may not be willing to go quite that far; however, a probability of 0.9999 would be reasonable. This is the same as saying only 1 failure is expected in 10,000 events. This failure is eventually expected due to the life of any one of the many variables that must function in order for the light to operate.

Variables that MUST function properly include: bulb, socket, wire to socket, switch, wire to switch, breaker, wire to breaker box, meter, wire to meter, transformer, wire to transformer, substation, wire to substation, power plant generator, and the list goes on…..

Each of these vital variables has subcomponents that must function properly in order for the system to function. As we get further away from the light bulb, things get more complicated. The power generation plant has thousands of subcomponents and raw material requirements.

With all of these must-function players, why can we say the probability of the light coming on when the switch is flipped to 'ON' is near 1? The answer can be reduced to one word, i.e., Design. The process has been engineered, built, and maintained to ensure close to 100% functionality.

Now, ask the probability of the light coming on without the benefit of design. With no design, all these individual components would have to be in place by coincidence. This means you could go where no man has gone before and find all this in place and functioning. As you journey through uninhabited and untraveled wilderness, you come upon a house that just evolved and inside the front door there is a light switch. No one built this; it just evolved. Once inside you flip the switch, and the light by the front door comes on. This indicates that all of the pieces are in place back to and beyond the power generation plant. This all happened without design. It just fell together through random chance. Think about it. All of the raw materials are available in or on Earth. Why couldn't they just come together?

The design requirement is easily seen in this example, yet in some of the most brilliant minds, the far more complex construction of planet Earth and humankind is plausible without design. If you still believe we are the product of random chance, take a walk in the wilderness; you may find a coincidental functioning light bulb.

Ann and Larry

Me and Ordinary Miracles

BIO: Larry Austin Henderson, Sr. lives in Lucedale, Mississippi with spouse, Elizabeth Ann Baygents Henderson. They have two children who are practicing physicians, and five grandchildren who are adorable. Larry is a Chemist, a retired Quality Director, an active bicyclist, gardener, and boater.

Why Ordinary Miracles:
Throughout my life, I have been interested in science and fascinated by life. I was raised a Christian and still routinely attend protestant church. Ordinary Miracles began in a church service during a segment called ***Praise***

Reports where the congregation is invited to share blessings that they have received during the previous week. This had gotten somewhat predictable with respect to what was going to be offered as things that fellow members were thankful for: 'I am thankful for the beautiful weather that we are having,' 'I was blessed with an accident free trip to Mobile,' or 'I am thankful for a successful business venture.' After everyone that wanted to offer praise had spoken, I said, "I am thankful for Ordinary Miracles" and told about the ozone layer that protects life on Earth's surface from harmful gamma radiation. Each Sunday thereafter, I had another, and there are more than a few. I wrote them down to avoid repetition. Once the notes got bigger, they took on a life of their own.

www.ingramcontent.com/pod-product-compliance
Lightning Source LLC
Chambersburg PA
CBHW051709170526
45167CB00002B/594